特别鸣谢
国家林业和草原局
成都市科学技术局
对本书出版提供的支持

Special thanks to
the National Forestry and Grassland Administration
and the Chengdu Science and Technology Bureau
for supporting the publication of this book

成都大熊猫繁育研究基地
CHENGDU RESEARCH BASE OF GIANT PANDA BREEDING 1987

成都大熊猫繁育研究基地　版权所有

未经版权所有方许可，不得对本书内容进行转载、翻印，或以其他手段进行传播。

Copyright © 2023
Chengdu Research Base of Giant Panda Breeding

All rights reserved. No part of this publication maybe reproduced, stored in a retrieval system, or transmitted in any form or by any means, electronic, photocopying, recording, or otherwise, without the prior written permission of the copyright owner.

你不知道的
THE GIANT PANDA YOU DON'T KNOW
大熊猫

金双 / 著

JIN SHUANG

四川少年儿童出版社

图书在版编目（CIP）数据

你不知道的大熊猫 / 金双著 . — 成都 : 四川少年儿童出版社 , 2022.9（2023.3 重印）
ISBN 978-7-5728-0877-7

Ⅰ . ①你… Ⅱ . ①金… Ⅲ . ①大熊猫 – 普及读物
Ⅳ . ① Q959.838-49

中国版本图书馆 CIP 数据核字 (2022) 第 151220 号

NI BU ZHIDAO DE DAXIONGMAO
你不知道的大熊猫

金 双 / 著

出 版 人	常 青
责任编辑	李明颖
美术编辑	张 雪
装帧设计	良友设计
插画绘制	ENO 傅义涵 郎佳琪
责任校对	张舒平
责任印制	王 春
出 版	四川少年儿童出版社
地 址	成都市锦江区三色路 238 号
电 话	028-86361727（发行部）
网 址	http://www.sccph.com.cn
网 店	http://scsnetcbs.tmall.com
经 销	新华书店
印 刷	成都兴怡包装装潢有限公司
成品尺寸	240mm×185mm
开 本	16
印 张	14
版 次	2023 年 1 月第 1 版
印 次	2023 年 3 月第 2 次印刷
书 号	ISBN 978-7-5728-0877-7
定 价	58.00 元

文学指导
Literary Supervisor
秦丽、马佳

Qin Li, Ma Jia

科学顾问
Scientific Advisor
侯蓉、黄祥明、吴孔菊、沈富军、
唐亚飞、向波、许萍、杨奎兴、陈欣

Hou Rong, Huang Xiangming,
Wu Kongju, Shen Fujun,
Tang Yafei, Xiang Bo,
Xu Ping, Yang Kuixing, Chen Xin

英文审订
English Revision
吴樱

Wu Ying

图片提供
Picture Provider
成都大熊猫繁育研究基地、
pandapia、张志和、崔凯、雍严格、
兰景超、廖骏、魏辅文、佘轶

Chengdu Research Base of Giant Panda Breeding,
pandapia, Zhang Zhihe,
Cui Kai, Yong Yan'ge, Lan Jingchao,
Liao Jun, Wei Fuwen, She Yi

你认识的大熊猫是什么样子的？
What do you know about giant pandas?

THE GIANT PANDA YOU DON'T KNOW

画出你印象中的大熊猫。
Picture the giant panda in your mind.

你不知道的大熊猫

序

在那么多种珍稀动物中,为什么我们特别关注大熊猫?因为,它是兼具极高社会价值和生态价值的完美物种。

1869年,法国人阿尔芒·戴维发现了大熊猫并为之命名,西方探险家开始追捧这一物种。他们非法收集大熊猫皮毛,猎杀大熊猫,甚至试图将大熊猫活体带走,让这本就濒危的物种生存愈加艰难。中华人民共和国成立后,大熊猫得到国家的保护,不仅在国内受到大众的广泛喜爱,还作为"和平使者"去到很多国家,在促进中外人民友谊的同时,进一步传播和推广了国际生物多样性保护的理念。

PREFACE

Among so many rare species, why are we paying so much attention to the giant panda? Because it is a significant species with extremely high social and ecological values.

In 1869, the Frenchman Armand David discovered and named the giant panda, setting off a panda craze among Western explorers. They poached giant pandas, collected their hides, and even tried to take them out of their homeland alive, making it harder for the endangered species to survive. Fortunately, the giant panda was later put under state protection and has not only won over the public at home, but also garnered massive popularity when traveling abroad as a "messenger of peace", further spreading and promoting the concept of international biodiversity conservation while strengthening the friendship of people between China and other countries.

谁看都萌的大熊猫自带明星光环。因为大熊猫，我们更加广泛地唤起了大众对物种保护行动的关注。保护大熊猫，保护它们的栖息地，也同时连带性地保护了诸如川金丝猴、羚牛、横斑锦蛇、独叶草、岷江冷杉等大量容易被忽视的物种。关注大熊猫，其实关注的是整个生态环境。

科学发现大熊猫150年来，中国大熊猫研究团队从野外到实验室，从国内独立研究到与国际机构合作，对大熊猫的生存和发展不断进行探索，取得了一项又一项研究成果，大熊猫这一物种的神秘面纱也逐渐被揭开。这本书将大熊猫科学研究的成果，大熊猫鲜为人知的秘密和它们经常被忽略的特点等用通俗易懂的方式呈现出来，让我们发现大熊猫原来是一种这么有故事的动物，原来我们对它们的一些行为存在误解，原来我们可以为大熊猫保护做那么多力所能及的事情……

了解大熊猫越多，了解我们每个人赖以生存的大自然就越深入。保护大熊猫，你我同行！

张志和 博士、研究员

Thanks to the adorable giant panda, which is a star with massive appeal, we have succeeded in drawing the public's attention to species conservation. Protecting giant pandas and their habitats provide conservation for numerous other neglected species, including the golden snub-nosed monkey, takin, *Elaphe perlacea, Kingdonia uniflora*, and *Abies faxoniana*. The attention given to giant pandas extends to the entire ecological environment.

Over the past 150 years since the scientific discovery of the giant panda, China's giant panda research team has been exploring the giant panda from the wild to the laboratory and from domestic coordination to international cooperation, achieving laudable research results one after another and gradually unveiling the mysteries of the giant panda. This book presents the scientific research achievements, little-known secrets, and often-overlooked features of giant pandas in an easy-to-understand manner, enabling us to learn about many stories behind them and discover that we misunderstand their behaviors, and we can do a lot with our power to protect them...

The more we learn about giant pandas, the deeper we understand the earth on which each of us lives. Come with us and learn more about the giant panda.

Professor Zhang Zhihe, Ph.D.

目录

前言   16
身份证   20

NO.1   大熊猫的"老祖宗"   22
NO.2   大熊猫名字的由来   38
NO.3   史上最强动物外交家   52
NO.4   大熊猫的"孤岛"困境   64
NO.5   和"团子"做邻居是啥感觉   76

NO.6.   大熊猫会冬眠吗   88
NO.7   大、小熊猫可别傻傻分不清   96
NO.8   特殊食谱   106
NO.9   神奇的伪拇指   116
NO.10   牙齿那点儿事   122

CONTENTS

Introduction 17
ID Card 21

NO.1 Giant Panda's Ancestors 22
NO.2 The Origin of the Giant Panda's Name 38
NO.3 The Most Charming Animal Diplomat in History ... 52
NO.4 Giant Panda's Habitat Fragmentation Plight 64
NO.5 What's It Like to Be a Giant Panda Neighbor 76

NO.6 Do Giant Pandas Hibernate 88
NO.7 Distinction Between Giant Pandas and Red Pandas ... 96
NO.8 Unique Diet 106
NO.9 Magical Pseudo-thumb 116
NO.10 Giant Panda's Teeth 122

NO.11 吃竹子的食肉目动物　　130
NO.12 我"懒"我有理　　138
NO.13 熊猫便便是香的　　144
NO.14 被忽略的尾巴　　150
NO.15 大熊猫的"信号源"　　156

NO.16 "滚滚"成长记　　162
NO.17 大熊猫小满的故事　　178
NO.18 教你听懂"熊猫语"　　186
NO.19 难缠的疾病　　192
NO.20 来做大熊猫小卫士　　208

致谢　　220
成都大熊猫繁育研究基地　　224

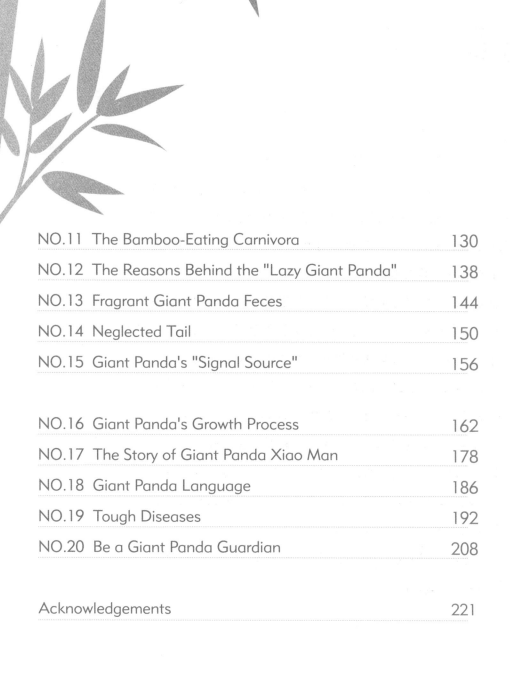

NO.11 The Bamboo-Eating Carnivora   130

NO.12 The Reasons Behind the "Lazy Giant Panda"   138

NO.13 Fragrant Giant Panda Feces   144

NO.14 Neglected Tail   150

NO.15 Giant Panda's "Signal Source"   156

NO.16 Giant Panda's Growth Process   162

NO.17 The Story of Giant Panda Xiao Man   178

NO.18 Giant Panda Language   186

NO.19 Tough Diseases   192

NO.20 Be a Giant Panda Guardian   208

Acknowledgements   221

前言

网上有一段只有57秒的视频，视频里一只叫"奇一"的大熊猫宝宝抱着"奶爸"（饲养员）的大腿不松手。这个视频刚放到网上一天，播放量就轻松过万。提起超级网红大熊猫，那真叫无人不知、无人不晓，连世界自然基金会（WWF）的官方标志都是我们可爱的大熊猫呢！

可能你知道大熊猫是中国的国宝，知道萌化人心的它其实是猛兽，知道它喜欢吃竹子，知道它的网名是"滚滚"和"团子"，但是，身为动物界顶级流量明星的它，怎么可能就这点儿话题？

你知道最早的大熊猫可能有"欧洲血统"吗？你知道在中国古代战争中高举大熊猫旗是什么意思吗？你想要一个大熊猫版的吃竹子教程吗？你知道雄性大熊猫为什么要倒立着尿尿吗……

本书将带你走进大熊猫的世界，探索很多关于大熊猫的奥秘，让你了解扛过了冰川纪大灭绝、又猛又彪的大熊猫现在面临的威胁，以及如何成为一名大熊猫小卫士。

One, two, three, go! 你将遇见一只熟悉又陌生、故事多多的大熊猫。

INTRODUCTION

A 57-second video of a giant panda cub named Qi Yi hugging her keeper's leg effortlessly hit 10,000 views in just one day. The online sensation panda is well-known by the public and is even the official logo for the World Wide Fund for Nature (WWF).

Maybe you already know that the giant panda is China's national treasure and its cute facade and bamboo diet belie a fierce beast that has been dubbed Gungun and Tuanzi by Chinese netizens. However, as the animal kingdom superstar, how could it possibly be limited to just a few topics?

Do you know that the earliest giant panda may have European ancestry? Do you know what it meant to hold high giant panda patterned flags during war in ancient China? Do you want a bamboo-eating tutorial from the giant panda? Do you know why male pandas stand on their heads to discharge urine...

This book invites you into the world of giant pandas to discover their secrets and better understand the current threats the fierce and sturdy giant pandas that survived the Ice Age extinction are facing; and helps you to become a giant panda guardian.

One, two, three, go! You will meet a familiar, yet strange giant panda with many stories.

身份证

姓　名：大熊猫
别　名：貔貅、貘、驺虞、猫熊、黑白熊、花熊、竹熊、食铁兽
学　名：*Ailuropoda melanoleuca*
英文名：Giant Panda
出生地：**中国**

界：动物界　　　　　　目：食肉目
门：脊索动物门　　　　科：熊科
亚门：脊椎动物亚门　　亚科：大熊猫亚科
纲：哺乳纲　　　　　　属：大熊猫属
亚纲：真兽亚纲　　　　种：大熊猫

世界自然保护联盟濒危物种红色名录（IUCN）：易危（VU）
国家重点保护野生动物名录：国家一级保护动物

ID CARD

Name: Giant Panda
Nicknames: Pi Xiu, Mo, Zou Yu, Mao Xiong, black and white bear, mottled bear, bamboo bear, and iron-eater
Scientific name: *Ailuropoda melanoleuca*
English name: Giant panda
Place of origin: China

Kingdom: Animalia
Phylum: Chordata
Subphylum: Vertebrata
Class: Mammalia
Subclass: Eutheria
Order: Carnivora
Family: Ursidae
Subfamily: Ailuropodinae
Genus: *Ailuropoda*
Species: *Ailuropoda melanoleuca*

IUCN Red List of Threatened Species: Vulnerable (VU)
List of Wildlife under Special State Protection: Animal under first-class state protection

NO.1 The Giant Panda You Don't Know

Giant Ancestors

大熊猫的"老祖宗"

大熊猫和恐龙谁更古老？

你知道大熊猫为什么被称为"活化石"吗？

你相信大熊猫起源于欧洲吗？

你知道中国最早的大熊猫和现在的大熊猫长得并不像吗？

Panda's

Which one has survived on Earth longer: the giant panda or the dinosaur? Do you know why the giant panda is called a "living fossil"? Do you think that the giant panda originated in Europe? Do you know that the earliest giant panda ancestors in China looked quite different from present-day giant pandas?

大熊猫为什么被称为"活化石"?

WHY IS THE GIANT PANDA CALLED A "LIVING FOSSIL"?

恐龙出现在2亿3000万年前(三叠纪),大约在6500万年前(白垩纪晚期)就灭绝了。

最早的大熊猫出现在800万年前,虽然是在恐龙灭绝约5000万年后才出现的,但是也非常古老了。正因为古老并一直生存到现在,而且从大约200万年前到现代变化很小,所以它们被称为"活化石"。

Dinosaurs first appeared in the Triassic period 230 million years ago, and died out in the late Cretaceous period about 65 million years ago.

Although the earliest giant panda did not appear until about 8 million years ago after 50 million years of the extinction of the dinosaurs, the giant panda is pretty ancient! It is called a "living fossil" because it has survived from prehistoric times to the present and has barely changed for about 2 million years.

巴氏大熊猫下颌骨化石模型
Mandible fossil replica of *Ailuropoda baconi*

大熊猫起源于欧洲？

THE GIANT PANDA ORIGINATED IN EUROPE?

大熊猫是中国独有的吗？关于大熊猫的起源地究竟在哪里，一直存在着争议。

早在1942年，科学家就在匈牙利发现了700万年前的葛氏郊熊猫的牙齿化石。当时，很多学者都将葛氏郊熊猫看作是东亚大熊猫的祖先（后来发现它只是始熊猫的一个分支，已经灭绝了，没有留下后代）。

Are giant pandas exclusive to China? The origin of the giant panda remains a matter of debate.

As early as in 1942, scientists discovered fossilized teeth of the *Agriarcros goaci* (an ancient panda from about 7 million years ago) in Hungary, which was believed by many scholars at that time to be the ancestor of the giant pandas in East Asia (It was later confirmed to be an extinct branch of the *Ailurarctos lufengensis*).

After that, China successively discovered earlier giant panda fossils, constantly challenging the theory that giant pandas originated in Europe. From 1981 to 1983, researchers discovered giant panda fossils dating back 8 million

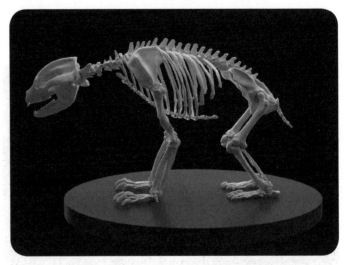

始熊猫骨架复原模型
Skeleton restoration replica of *Ailurarctos lufengensis*

在这之后，中国接连发现了年代更早的大熊猫化石，不断挑战着大熊猫欧洲起源论。1981年至1983年，研究人员分别在云南禄丰盆地和元谋盆地发现了800万年前的大熊猫化石，将其定名为"始熊猫"。这一发现为大熊猫的起源提供了直接证据。大熊猫起源于中国云南，这个结论被国内外学者认可了。

然而，2015年，科学家在西班牙发现了约1160万年前的大熊猫额骨和牙齿化石，大熊猫的起源又一次变得扑朔迷离。2017年，人们又在匈牙利发现了1000万年前的大熊猫牙齿化石，大熊猫欧洲起源论再起波澜。

但是，仅凭欧洲发现的这一两处化石就认定大熊猫起源于欧洲还为时尚早。也许，将来人们会在其他地方发现更古老的大熊猫化石。我们仍需要更多的发现来证明大熊猫究竟起源于哪里。

years ago in Lufeng Basin and Yuanmou Basin, Yunnan province, and named it *Ailurarctos lufengensis*. This discovery provides evidence proving the origin of the giant panda and concludes that giant pandas originated in Yunnan, China, which has been accepted by scholars around the world.

However, in 2015, scientists discovered fossilized frontal bones and teeth of giant panda about 11.6 million years ago in Spain, again putting the origin of the giant panda in doubt. In 2017, the discovery of a set of fossilized giant panda teeth dating back 10 million years ago in Hungary led to the resurfacing of the European origin theory for giant pandas.

It is too early to conclude that the giant panda originated in Europe from two fossil discoveries in Europe. Maybe in a few years, more ancient panda fossils will be found elsewhere. More evidence is needed to determine from where the giant panda originated.

武陵山大熊猫右上颌骨化石模型
Right maxilla fossil replica of *Ailuropoda wulingshanensis*

THE "ANCESTORS" OF GIANT PANDAS IN CHINA

中国大熊猫的"老祖宗"们

The giant panda fossils discovered in China have been recognized as the direct ancestor to the present-day giant panda. Let's take a look at the progenitors of the Chinese giant panda!

在中国陆续发现的大熊猫化石，目前已经被公认是现生大熊猫的直系祖先。我们来看看中国大熊猫的"老祖宗"们吧！

历史上大熊猫分布范围的大小变化
Changes in Distribution Ranges for Giant Pandas in History

历史上，大熊猫曾经分布广泛，但近代以来，它们的野外栖息地急剧缩小。

Giant pandas were once widely distributed in history, but their wild habitat has shrunk dramatically in the modern era.

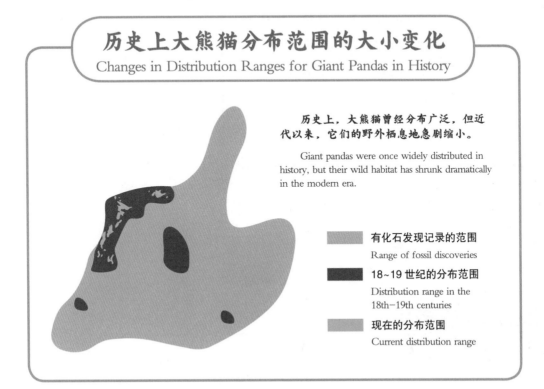

- 有化石发现记录的范围 / Range of fossil discoveries
- 18~19 世纪的分布范围 / Distribution range in the 18th–19th centuries
- 现在的分布范围 / Current distribution range

800 万年前

↓

始熊猫

> 我是拟熊类演变成的早期大熊猫，住在温暖湿润的森林地带。我的臼齿（就是嘴里的大牙）很小，表面还有褶皱；我个子不高，块头和一条胖胖的狗差不多。

始熊猫（复原图）
Ailurarctos lufengensis (restored image)

8 million years ago
Ailurarctos lufengensis

"I am a primal panda evolving from a bear-like species and live in warm and humid forests. I have small molars with a grooved surface and a short body about the size of a plump dog."

NO.1 Giant Panda's Ancestors NO.1 大熊猫的"老祖宗"

2.5 million years ago
Ailuropoda microta

"Compared to the *Ailurarctos lufengensis*, I have a much bigger body and coarser and wider molars (boo-yah, my molars are bigger), so I am omnivorous and consume both bamboo and meat."

250 万年前

小种大熊猫

我比"始熊猫"的块头大多了，有比较粗糙宽大的臼齿（哈哈，我的大牙比较大）。我是又吃竹子又吃肉的杂食动物。

小种大熊猫（复原图）
Ailuropoda microta (restored image)

180 万年前

武陵山大熊猫

> 我比"小种大熊猫"的块头还要大一点儿！竹子已经是我的主食啦！

武陵山大熊猫（复原图）
Ailuropoda wulingshanensis
(restored image)

1.8 million years ago
Ailuropoda wulingshanensis

"I am even a bit larger than *Ailuropoda microta*! Bamboo is already a staple food in my diet！"

NO.1 Giant Panda's Ancestors   NO.1 大熊猫的"老祖宗"

750,000 years ago
Ailuropoda baconi

"I am the largest and most numerous, representing the peak in giant panda development. My appearance, behavior, and diet are nearly the same as my offsprings (*Ailuropoda melanoleuca*). I inhabit forests in warm and humid monsoon regions of the Yellow River, the ChangJiang River, and the Pearl River and even in Vietnam and northern Myanmar. My neighbors include *stegodon*, *smilodon*."

75 万年前

巴氏大熊猫

我块头最大，家族成员也最多，我生活在我们大熊猫家族最繁盛的时期。我的外貌、行为、吃的食物都和我的后代（现生大熊猫）差不多。我住在黄河、长江和珠江流域温暖湿润的季风林中，在越南和缅甸北部也能看到我的身影。我和剑齿象、剑齿虎等动物生活在一起。

巴氏大熊猫（复原图）
Ailuropoda baconi
(restored image)

1.5 万年前

↓

现生大熊猫

> 我没有"巴氏大熊猫"块头大，但是我的牙齿、咀嚼肌和凸出的矢状脊等结构都能让我轻松地咬碎坚硬的竹竿。虽然我的主食是竹子，但是我偶尔也吃一点儿动物尸体或其他植物。

现生大熊猫
Ailuropoda melanoleuca

15,000 years ago
Ailuropoda melanoleuca

"I'm not as big as *Ailuropoda baconi*, but my teeth, masticatory muscles, and protruding sagittal ridge give me the power to easily crush hard bamboo. Although my diet is dominated by bamboo, I occasionally eat animal carcasses or other plants."

NO.1 大熊猫的"老祖宗"

大熊猫体形大小的变化
The Body Size Change of the Giant Panda

始熊猫 *Ailurarctos lufengensis* (extinct) — 800 万年前 / 8 million years ago
小种大熊猫 *Ailuropoda microta* (extinct) — 250 万年前 / 2.5 million years ago
武陵山大熊猫 *Ailuropoda wulingshanensis* (extinct) — 180 万年前 / 1.8 million years ago
巴氏大熊猫 *Ailuropoda baconi* (extinct) — 75 万年前 / 750,000 years ago
现生大熊猫 *Ailuropoda melanoleuca* — 1.5 万年前 / 15,000 years ago

大熊猫的体形从古至今经历了从小变大再变小的过程。
From millions of years ago to the present, giant pandas have exhibited a rise and then decline in size.

大熊猫小百科

扛过冰川纪大灭绝的"活化石"

在距今270万年至12000年前的更新世，秦岭—淮河以南气候适宜、森林广布、食物丰富，哺乳动物繁盛，动物数量非常多。这里繁衍着一个动物群，生活着包括大熊猫、东方剑齿象、剑齿虎、巨猿、巨貘、猩猩、纳玛象、中国犀牛、水牛、猕猴、金丝猴、长臂猿、野猪、黑鹿、青羊、竹鼠、豪猪及亚洲黑熊等在内的众多动物族群。地质学家称该动物群为大熊猫—剑齿象动物群。

由于第四纪冰川的反复侵袭以及造山运动造成的自然环境的剧烈变化，进入全新世以后，这个动物群的许多成员都销声匿迹了，就连剑齿象、剑齿虎、中国犀牛、巨猿、巨貘等威武雄壮的猛兽都未能逃脱灭绝的厄运，但大熊猫却神奇地生存了下来，成为举世闻名的"活化石"。

GIANT PANDA TIDBITS

The "Living Fossil" that Survived the Ice Age Extinction

During the Pleistocene about 2.7 million to 12,000 years ago, the south of the Qinling Mountains—Huaihe River featured a suitable climate, extensive forests, abundant food, and prosperous mammals as well as other animals. The fauna that thrived here included giant pandas, *stegodon orientalis*, *smilodon*, *gigantopithecus*, giant tapirs, *orangutans*, *palaeoloxodon namadicus*, Chinese rhinos, buffaloes, macaques, snub-nosed monkey, gibbons, wild boars, sambars, blue sheep, bamboo rats, porcupines, and Asian black bears and is referred to as the *Ailuropoda—Stegodon* Fauna by geologists.

Due to repeated attacks during the fourth glacial epoch and dramatic changes in the natural environment, such as the orogenic movement, many different species disappeared after entering the Holocene Epoch. Even ferocious beasts, including the *stegodon*, *smilodon*, Chinese rhino, *gigantopithecus*, and giant tapir failed to escape the fate of extinction, but the giant panda miraculously survived and became a world renowned "living fossil".

NO.2 The Giant Panda You Don't Know

The Origin the Giant Name

大熊猫名字的由来

你知道古代的貔貅和大熊猫是什么关系吗?
西晋时两军交战中举起大熊猫旗是什么意思呢?
为什么中国大熊猫的学名是外国人起的?

# of Panda's

Do you know the relationship between the ancient beast Pi Xiu and the giant panda? What did it mean to hold high giant panda patterned flags when two armies fought in a battle during the Jin Dynasty? Why did a foreigner give the scientific name for China's giant panda?

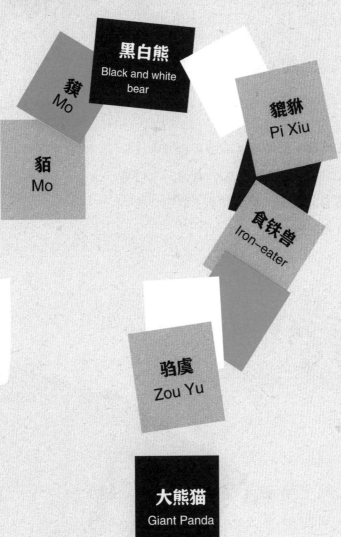

历史上对大熊猫的传奇记载

翻开历史长卷,我国古代关于大熊猫的传奇记载可真不少。

3000多年前西周初年汇编的《尚书》和《诗经》里,就出现了"貔貅(pí xiū)"这个名词。人们用"貔"比拟威武英勇的勇士;军队打着"貔""貅"的旗号,象征着战无不胜。

貔貅
Pi Xiu

THE LEGENDARY RECORDS OF THE GIANT PANDA IN HISTORY

Looking through historical scrolls from China, recorded legends about giant pandas abound.

As recorded in the *Book of Documents* and *Book of Songs*, more than three millennia ago in the early Western Zhou Dynasty, brave warriors were compared to the Pi, and the army even adopted Pi and Xiu patterned flags to symbolize invincibility.

NO.2 The Origin of the Giant Panda's Name   NO.2 大熊猫名字的由来

You can see the names of Pi Xiu, Mo, Zou Yu, Mao Xiong, black and white bear, mottled bear, bamboo bear, and iron-eater... all refer to giant pandas! Why did they have so many different names? It is because giant pandas were much more widely distributed in ancient times than they are today, and people in different regions gave them different names, so giant pandas possessed more than a dozen of names.

"如虎如貔如熊如黑"
——《尚书》中对大熊猫的描述
"Like a tiger, a pi, a bear, and a brown bear"
— Giant panda description in *Book of Documents*

你瞧，貔貅、貘、驺虞、猫熊、黑白熊、花熊、竹熊、食铁兽……指的都是大熊猫！为什么大熊猫有这么多不同的名字呢？那是因为在古代，大熊猫的分布范围比现在广得多，不同地方的人对大熊猫有不同的叫法，所以它们的名字有十几个呢！

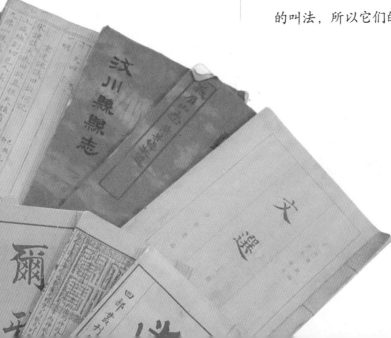

记载着与大熊猫有关史料的古籍
Ancient books that recorded historical facts about giant pandas

阿尔芒·戴维与大熊猫的故事

大熊猫在各地的名称，就像我们说的方言一样，只是在中国的某个地域内使用，而大熊猫的拉丁学名，是可以在全世界通用的名字。说到这里，我要介绍一位很重要的人物，他就是发现大熊猫并为之命名的法国神父皮埃尔·阿尔芒·戴维。

戴维出生在法国的一个山区小镇，从小聪慧好学、活泼好动，非常喜爱动植物。戴维青年时期立志要做一名科学家，到世界各地去考察动植物。

戴维手绘的大熊猫
Hand-drawn giant panda by Armand David

皮埃尔·阿尔芒·戴维（1826—1900）
Pére Armand David

THE STORY BETWEEN ARMAND DAVID AND GIANT PANDA

The various names for the giant panda, just like dialects, are limited to certain regions in China, while its scientific name is recognized all over the world. Speaking of which, I would like to introduce a very important person, the French Priest Pére Armand David, who discovered and named the giant panda.

Born in a small town in a mountainous area of France, Armand David was smart, studious, and lively. He developed a keen interest in animals and plants from an early

19世纪的法国马赛港，皮埃尔·阿尔芒·戴维从这里出发前往中国
Port of Marseille, France in the 19th century, from where Pére Armand David began his journey to China

age. In his youth, David was determined to become a scientist and travel around the world to investigate plants and animals.

He was also a devout Catholic, who went to a Catholic institution to study at 20 and became a priest at 24.

In 1862, Priest David arrived in China as a missionary. In addition to preach, he shouldered another mission, which was to collect animal and plant specimens for the Muséum National d'Histoire Naturelle.

When David learned that Baoxing of Sichuan boasted rich biodiversity and abundant animals and plants, some of which were rare species rarely seen by people, he headed for Baoxing from

他也是一个虔诚的天主教徒，20岁时到天主教培训机构去学习，24岁就当上了神父。

1862年，戴维神父来到中国传教。除了传教，他还肩负着另外一项使命，那就是为法国自然历史博物馆收集动植物标本。

在得知四川宝兴一带生物多样性丰富，动植物种类繁多，其中有一些是人们难得一见的珍稀物种的消息后，戴维从上海来到宝兴，担任了宝兴邓池沟教堂的第四任神父。

邓池沟教堂俯瞰
Aerial view of Dengchigou Church

1869年的春天，戴维于考察途中在一个姓李的猎人家歇脚时，第一次看到了挂在墙上的一张毛色黑白相间的兽皮，这引起了他极大的兴趣。根据兽皮的形态特征，他认为这是一种珍稀的熊类，是一个新物种，于是给它起了个学名叫 Ursus melanoleucus（在拉丁文里的意思是黑白相间的熊）。

Shanghai to serve as the fourth priest of the Dengchigou Catholic Church, Baoxing.

In the spring of 1869, David was passing through and stopped at a house owned by a hunter surnamed Li for a break on the way to conduct a field investigation and noticed a black and white hide hanging on the wall for the first time, which piqued his curiosity. According to the morphological characteristics of the hide, he discovered it was a rare *Ursus* genus and a new species, so he named it *Ursus melanoleucus* (Latin for black and white bear).

邓池沟教堂内景
Interior view of Dengchigou Church

THE SCIENTIFIC NAME UP THE GIANTI PANDA

In 1870, Alphonse Milne Edwards, director of the Muséum National d'Histoire Naturelle, studied specimens from the "black and white bear" that David had sent to France.

Edwards opined that the "black and white bear" is more similar to a red panda found in Tibet, China than a bear because they share many similarities despite the large disparity in body size and changed its genus from *Ursus* to *Ailuropoda* but retained the species name *melanoleuca*, so it became *Ailuropoda melanoleuca* (black and white panda).

大熊猫的科学命名

1870年，法国自然历史博物馆馆长，阿方斯·米尔恩·爱德华兹对戴维神父寄回法国的"黑白熊"标本重新进行了研究。

爱德华兹认为"黑白熊"更接近之前在中国西藏发现的小熊猫，认为"黑白熊"和小熊猫有很多相似之处，只是体形差别有点儿大。所以，爱德华兹把戴维为黑白熊起的属名*Ursus*（熊属）改为*Ailuropoda*（猫熊属），种名*melanoleuca*保留，组成了*Ailuropoda melanoleuca*（黑白相间的猫熊）这个名字。

法国自然历史博物馆，戴维神父寄回法国的大熊猫标本就保存在这里
Muséum National d'Histoire Naturelle, which housed the giant panda specimens that Father David sent to France

后来，关于大熊猫的学名，分类学家们经过反复研究讨论，认为Ailuropoda melanoleuca（黑白相间的猫熊）最合适，这一学名也就一直被使用并延续了下来。这就是大熊猫学名的来历。

在戴维发现大熊猫之前，大熊猫已经有了十几个名字，应该说中国人早就发现了大熊猫，为什么要把发现大熊猫的功劳给戴维呢？

因为，中国人早期给大熊猫起的名字大多是根据外貌、体毛颜色、方言来的，而戴维是第一个按照国际公认的标准给大熊猫科学命名的人，所以戴维才是科学发现大熊猫并把这一稀世珍宝介绍给全世界的第一人。

戴维在中国12年，共发现了189个新物种，包括大熊猫、金丝猴、羚牛、麋鹿、珙桐（鸽子树）、大叶杜鹃、报春花等。

Later, taxonomists acknowledged that *Ailuropoda melanoleuca* was the most appropriate scientific name for the giant panda after repeated research and discussions, so the name was adopted and is still in use today. This is the origin of the giant panda's scientific name.

Since the Chinese had already given the giant panda more than a dozen of names before David's discovery, we can conclude that the Chinese had discovered the giant panda long ago, so why the credit for giant panda discovery should be given to David?

In fact, the earliest Chinese names for the giant panda were given primarily based on its appearance and fur color or through local dialects, while David was the first to name it scientifically according to internationally recognized standards. Therefore, David was the first person to discover the giant panda in a scientific sense and introduced this rare treasure to the world.

Armand David discovered a total of 189 new species during his 12-year stint in China, including the giant panda, golden snub-nosed monkey, takin, Père David's deer, Chinese dove tree (*Davidia involucrata*), *Rhododendron faberisp*, *Primula malacoides*, and so much more.

"CAT BEAR" OR "BEAR CAT"?

In English, the giant panda is also known as the Cat Bear (Mao Xiong in Chinese) because its face resembles a cat and the body is that of a bear.

In August 1939, China exhibited a captive giant panda for the first time in the Chongqing Beibei Civilian Park (present-day Beibei Park). The giant panda's nametag at that time included both the Chinese and English names, each on its own row. In keeping with the writing standards of the English name in the upper row, the Chinese name underneath was also written from left to right.

However, Chinese at that time was written from right to left, so visitors called it Xiong Mao instead of Mao Xiong. Hereafter, the name Da Xiongmao was established and has been used since.

"猫熊"还是"熊猫"?

英语里,人们称大熊猫为"Cat Bear",中文翻译过来是"猫熊",因为它的脸像猫,身体像熊。

1939年8月,在重庆北碚平民公园(现在叫北碚公园),大熊猫第一次在国内展出。展出的标牌上分别用中文、英文书写了大熊猫的名字:上面一排是英文名字,下面一排的中文名字也按照英文的书写方式,从左到右写上了"猫熊"二字。

由于当时中文的书写和阅读习惯是从右至左,所以游人都将"猫熊"读成了"熊猫"。此后,"大熊猫"这一称呼便约定俗成,从此流传了下来。

FUN FACTS

趣闻一串串

黄帝战炎帝 大熊猫来助阵

司马迁在《史记·五帝本纪》中记载，远在4000年前，有一个部落的首领叫黄帝，他利用驯养的虎、豹、熊、貔貅（大熊猫）等猛兽助战，在阪泉（今河南涿鹿县）打败了另一个部落首领炎帝。

Giant Pandas Aided the Yellow Emperor to Defeat the Yan Emperor

In the *Biography of Five Emperors in Historical Records*, Sima Qian recorded that as far back as four millennia ago, a tribal leader known as the Yellow Emperor utilized domesticated beasts like tigers, leopards, bears, and Pi Xiu (giant pandas) to assist him in battle and defeated another tribal leader Yan Emperor in Banquan (present-day Zhuolu County, Henan province).

阪泉之战
The war of Banquan

FUN FACTS

趣闻一串串

认怂了？举起大熊猫旗吧！

西晋时（1700年前），人们称大熊猫为驺虞（zōu yú）。因为它只吃竹子，不伤害和猎食其他动物，是一种能与友邻和平共处的"义兽"，所以当时的人们把大熊猫看作和平友好的象征。当两军交战，杀得天昏地暗、日月无光时，只要有一方举起驺虞旗，战斗就会戛然而止，因为当时的战争规则是：举起驺虞旗，就表示请求休战，停止冲突。

Do You Want a Truce? Raise Your Panda Flag!

In the Western Jin Dynasty (1700 years ago), the giant panda was called Zou Yu. Since they only ate bamboo and rarely harmed or preyed on other animals, they were deemed a "righteous beast" that could live in peace with its neighbors and regarded as a symbol for peace and friendship at that time. During a battle where two armies fought with everything they had, as long as one side raised the Zou Yu (giant panda) flag, the battle would immediately cease. As the rules of war stated: Any use of the Zou Yu flag indicates a request for a truce and an end to the fighting.

NO.2 The Origin of the Giant Panda's Name　　NO.2 大熊猫名字的由来

驺虞示和
Reconcilement of Zou Yu

The Giant Panda You Don't Know

NO.3

The Most Animal in History

史上最强动物外交家

大熊猫作为对外交流大使最早可以追溯到哪个朝代?
你知道大熊猫出国的三种途径吗?

Charming Diplomat

To which dynasty can panda diplomacy date back? Do you know the three ways giant pandas can go abroad?

大熊猫对外交流源远流长

  古时候大熊猫被看成是威猛的图腾，还常常被当作稀世珍宝敬献给帝王。如果打仗时举起大熊猫旗，战争马上就能停止，真是很神奇！

  正是因为大熊猫非常珍贵，又是和平的象征，所以从古至今，它们多次被送到国外，肩负着对外交流使命。这个"动物外交家"做得很棒呢！

  据日本《皇家年鉴》记载，公元685年10月22日，大唐女皇武则天送给日本天武天皇两只活体白熊（大熊猫）和70张白熊毛皮，用以表示友好和慷慨。这是大熊猫在国际交往中担任"交流使者"的最早记录。

  1957年到1983年，中华人民共和国先后向苏联、朝鲜、美国、日本、法国、英国、西德、墨西哥和西班牙9个国家赠送了24只大熊猫，这些"交流使者"受到了各国人民的极大欢迎，对发展我国的对外友好关系做出了巨大贡献。

THE GIANT PANDA DIPLOMACY WITH A LONG HISTORY

  In ancient times, the giant panda was utilized as the totem of power and was often offered to the emperor as a rare treasure. It is amazing that the giant panda patterned flag could immediately suspend battle.

  It is because giant pandas are so precious and symbolize peace that they have been sent abroad on diplomatic missions since ancient times and have done well as "animal diplomats"!

  According to the Japanese *Royal Yearbook*, on October 22, 685 AD, Empress Wu Zetian of the Tang Dynasty presented two live white bears (giant pandas) and 70 hides to Emperor Tenmu of Japan as a gesture of amicability and generosity, which is the earliest record of giant pandas as "diplomatic envoys" in international exchange.

  From 1957 to 1983, the People's Republic of China successively gifted 24 giant pandas to nine countries: the Soviet Union, North Korea, the United States, Japan, France, the United Kingdom, Federal Republic of Germany, Mexico, and Spain. The public in every country they visited warmly welcome these "diplomatic envoys", significantly contributing to the development of friendly relations with foreign countries.

大熊猫在海外受到了人们的热情追捧
Giant pandas were widely popular overseas

中外友谊的桥梁

到了1983年，出于保护濒危动物的需要，我国取消了向国外无偿赠送大熊猫的方式。大熊猫作为国礼的时代结束了，但是大熊猫的出国之路并没有因此中断。

此后，许多国家纷纷采用短期借展的方式请大熊猫到当地动物园做客。截至1992年12月，10年间共有50多只中国大熊猫前往亚洲、非洲、欧洲、美洲、大洋洲的10多个国家、30余座城市巡展。它们每到一处，都受到该国政府和人民的热烈欢迎，增强了中国人民与世界人民之间的友谊。

A BRIDGE OF FRIENDSHIP BETWEEN CHINA AND FOREIGN COUNTRIES

However, in 1983, to protect endangered animals, China terminated its policy of giving pandas away to foreign countries, ending the era of giant pandas as a national gift, while continuing along the road of giant pandas going abroad in other ways.

Since then, many countries had adopted short-term loan exhibitions for giant pandas and invite them to their local zoos. By December 1992, more than 50 giant pandas had traveled to more than 30 cities across more than 10 countries over five continents: Asia, Africa, Europe, America, and Oceania. The government and people of that country excitedly welcome them, which had developed the friendship between the Chinese people and the people of the world.

联合国开发计划署首对动物大使——"启启"和"点点"
UNDP's first animal ambassadors - Qi Qi and Dian Dian

NO.3 The Most Charming Animal Diplomat in History NO.3 史上最强动物外交家

AMBASSADORS FOR SCIENTIFIC RESEARCH EXCHANGE

科研交流大使

Since 1993, commercial loan exhibitions had been abolished, and now the only way to go abroad for giant pandas is scientific cooperation and exchange.

Nowadays, giant pandas travel to foreign lands as "ambassadors for scientific research exchange" accompanied by technical personnel and normally stay abroad for a fixed term of 10 years. Panda

从1993年起，商业借展的方式也被取消了。如今，大熊猫出国只剩下科研合作交流这一途径。

大熊猫以"科研交流大使"的身份去到国外，我国也会派技术人员一同前往。一般来说，这些大熊猫的"留洋"期为10年，在国外产下的大熊猫宝宝也属于中国，而且大熊猫宝宝在3岁之前就要回到祖国。这期间如果大熊猫死亡，尸体也要归还给中国。

1999年，成都双流机场，大熊猫启程前往美国亚特兰大动物园
In 1999, giant pandas departing for the Atlanta Zoo, USA at Chengdu Shuangliu Airport

截至2020年11月，共有64只大熊猫以科研合作交流的方式生活在美国、俄罗斯、日本、西班牙、法国、加拿大、德国、奥地利、比利时、荷兰、英国、芬兰、丹麦、泰国、韩国、印度尼西亚、马来西亚、新加坡、澳大利亚等国家。有10只大熊猫生活在我国的香港、澳门、台湾地区。

还有2只生活在墨西哥的大熊猫，是中国政府以前赠送的，所以它们属于墨西哥。这两只大熊猫现在都已经30多岁了，相当于人类的90多岁高龄。现在，它们正在墨西哥安享晚年。

可爱的大熊猫征服了海外的无数民众，大熊猫成了"中国"的代名词。从喜爱大熊猫开始，一些人慢慢对中国和中国文化产生了兴趣。大熊猫用它独特的魅力，搭起了中国与世界友好交往的桥梁，把中国人民的友谊撒播到了世界各地。

cubs born abroad still belong to China and will return to China before they reach the age of 3. In addition, the panda's body will also be sent back to China in the event of death during this period.

As of November 2020, a total of 64 giant pandas are living in 20 countries, including the United States, Russia, Japan, Spain, France, Canada, Germany, Austria, Belgium, the Netherlands, the United Kingdom, Finland, Denmark, Thailand, South Korea, Indonesia, Malaysia, Singapore, Australia for scientific research cooperation and exchanges. There are 10 giant pandas living in Hong Kong, Macao, and Taiwan of China.

And there are two giant pandas living in Mexico were originally given away by China, so they belong to the country. Both two pandas are now in their 30s, which is equivalent to a nonagenarian, and are currently enjoying their twilight years in Mexico.

The lovely giant panda has won over millions of people around the world and has become synonymous with China. Beginning with their love of giant pandas, more people have gradually developed an interest in China and Chinese culture. By utilizing its unique charm, the giant panda has built a bridge of friendly exchanges between China and the rest of the world, spreading the seed of friendship all over the world.

国外动物园 / Foreign Zoos

国家	动物园	数量(只)	Country	Zoo	Quantity
奥地利	维也纳美泉宫动物园	2	Austria	Schoenbrunner Tiergarten GmbH	2
比利时	天堂公园	5	Belgium	Pairi Daize Park	5
丹麦	哥本哈根动物园	2	Denmark	Copenhagen Zoo	2
法国	博瓦勒动物园	3	France	Zoo Parc de Beauval	3
荷兰	欧维汉动物园	3	Netherlands	Ouwehands Zoo	3
西班牙	马德里动物园	3	Spain	Zoo Aquarium de Madrid	3
德国	柏林动物园	4	Germany	Zoo Berlin	4
英国	爱丁堡动物园	2	UK	Edinburgh Zoo-Scottish National Zoo	2
芬兰	艾赫泰里动物园	2	Finland	Zoo Ahtari	2
美国	华盛顿国家动物园	4	USA	National Zoological Park	4
美国	亚特兰大动物园	4	USA	Zoo Atlanta	4
美国	孟菲斯动物园	2	USA	Memphis Zoological Garden & Aquarium	2
加拿大	卡尔加里动物园	3	Canada	Calgary Zoo	3
墨西哥	查普尔特佩克动物园	2	Mexico	Zoologico de Chapultepec	2
俄罗斯	莫斯科动物园	2	Russia	Moscow Zoological Park	2
泰国	清迈动物园	1	Thailand	Chiang Mai Zoo	1
日本	神户动物园	1	Japan	Kobe Oji Zoo	1
日本	东京上野动物园	3	Japan	Ueno Zoological Gardens	3
日本	和歌山白浜野生动物园	6	Japan	Nanki Shirahama Adventure World	6
韩国	爱宝乐园	3	South Korea	Ever land Zoological Gardens	3
马来西亚	马来西亚动物园	3	Malaysia	Zoo Negara Malaysia	3
新加坡	新加坡动物园	2	Singapore	Singapore Zoological Gardens	2
澳大利亚	阿德莱德动物园	2	Australia	Adelaide Zoo	2
印度尼西亚	塔曼动物园	2	Indonesia	Taman Safari	2

港澳台地区动物园 / Zoos in Hong Kong, Macao, and Taiwan

地区	动物园	数量(只)	Region	Zoo	Quantity
香港	香港海洋公园	3	Hong Kong	Ocean Park Hong Kong	3
台湾	台北动物园	3	Taiwan	Taipei Zoo	3
澳门	石排湾郊野公园	4	Macao	Parque De Seac Pai Van	4

趣闻一串串

赴美 它让尼克松夫人惊声尖叫

　　1972年2月，尼克松总统在北京中南海与毛泽东主席握手，翻开了中美关系新的一页。在一次宴会上，尼克松夫人拿着熊猫牌香烟的盒子爱不释手地看，周恩来总理微笑着指着烟盒上的大熊猫图案对她说："您喜欢这个吗？你们把两只麝香牛送给中国人民，我们准备送两只大熊猫给美国人民。"尼克松夫人一听惊喜不已，对正与人谈话的尼克松惊声尖叫道："我的天哪！你听到总理说什么了吗？他要送大熊猫给我们！动物园会被挤垮的！"第二天，这一消息成为世界各大报刊的要闻！

　　1972年4月，来自四川宝兴县的大熊猫"玲玲"和"兴兴"乘专机从北京起飞，到达位于华盛顿的美国国家动物园，数千市民冒雨前来迎接。美国人民把1972年称为"大熊猫年"。这一年，"玲玲"和"兴兴"频频登上全美各大报刊的头版和封面，把世界性的第三次大熊猫热推向高潮。在这之后，每年有超过300万人次的游客来到华盛顿国家动物园参观大熊猫。

"玲玲"和"兴兴"在美国
Ling Ling and Xing Xing in the USA

FUN FACTS

Gifting a Pair of Pandas to the United States Made First Lady Pat Nixon Scream in Excitement

In February 1972, President Nixon shook hands with Chairman Mao Zedong at Zhongnanhai (headquarters of the central govenment) in Beijing, marking a new chapter in Sino-US relations. At a banquet, when Mrs. Nixon was holding a panda patterned cigarette case and observing it fondly, Premier Zhou Enlai pointed at the panda pattern on the cigarette case with a smile and said to her, "Do you like this? You gifted two musk oxen to the Chinese people, so we are going to give two giant pandas to the American people." Mrs. Nixon was so surprised that she excitedly said to Nixon, who was in the middle of a conversation, "Oh my God! Did you hear what the Prime Minister said? He is going to give us pandas! The zoo will be swarmed with visitors!" The next day, the gift made headlines in major newspapers all around the world!

In April 1972, giant pandas Ling Ling and Xing Xing from Baoxing, Sichuan departed from Beijing by special charter to the National Zoological Park in Washington and were welcomed by thousands of Americans, despite the downpour. Americans dubbed 1972 the Year of the Giant Panda. Ling Ling and Xing Xing frequently appeared on the front pages and covers of major American newspapers and magazines, culminating in the peak of the world's third panda fever. Since then, more than 3 million visitors came to National Zoological Park to see the giant pandas every year.

趣闻一串串

赴日 它享受高规格战斗机护航

　　1972年9月，日本首相田中角荣率团来北京进行建交谈判，其间表达了渴求大熊猫的愿望，毛泽东主席和周恩来总理答应了这一请求。

　　1972年10月，同样来自四川宝兴县的大熊猫"兰兰"和"康康"乘专机飞往日本。当飞机进入日本领空时，日本政府派战斗机升空护航。在东京机场，"兰兰"和"康康"受到了日本内阁大臣的亲自迎接。在警车护送下，它们乘车前往东京上野动物园，无数市民夹道欢迎。日本报纸称："大熊猫为中日友好史话书写了新篇章，日本人民为之神魂颠倒。"在上野动物园，每天有上万观众排队几小时只为一睹大熊猫的风采，动物园每年收到日本各地寄给大熊猫的贺信数千封。1979年4月，"兰兰"因病去世，动物园举行了数千人参加的追悼会，人们排着长队为"兰兰"送行。动物园收到的来自日本全国的唁电、鲜花数不胜数。日本内阁大臣在唁电中说："它是在为中日友好做出贡献之后死去的……"

日本小学生参观"兰兰"和"康康"
Japanese primary school students visiting Lan Lan and Kang Kang

FUN FACTS

The Panda Pair Gifted to Japan Given an Elite Fighter Escort

In September 1972, Japanese Prime Minister Kakuei Tanaka led a delegation to Beijing to negotiate on establishing diplomatic relations with China, during which he expressed his desire for giant pandas. Chairman Mao Zedong and Premier Zhou Enlai agreed to his request.

In October 1972, giant pandas Lan Lan and Kang Kang, also from Baoxing, Sichuan, flew to Japan by special charter. When the plane entered Japanese airspace, the Japanese government sent fighter jets to escort it. At Tokyo Airport, Lan Lan and Kang Kang were personally received by Japanese Cabinet Minister. On their way to Ueno Zoo, they were escorted by the police and warmly greeted by citizens. The Japanese newspaper remarked that the giant pandas have written a new chapter in the history of Sino-Japanese friendship, fascinating the Japanese people. Ueno Zoo received tens of thousands of visitors every day, who lined up for hours just to see the giant pandas, and thousands of letters welcoming them from across Japan every year. When Lan Lan died of illness in April 1979, the Zoo held a memorial service, which thousands of people who lined up to see Lan Lan off in addition to countless messages of condolences and flowers from all over Japan. A message of condolence from the Japanese Cabinet Minister stated: "It gave the ultimate sacrifice while contributing to Sino-Japanese friendship..."

The Giant Panda You Don't know

NO.4

Giant Habitat Plight

大熊猫的"孤岛"困境

人们是怎么对隐居山林的野生大熊猫进行"人口普查"的?
熬过了几百万年的大熊猫为什么在今天遭遇了生存危机?

Panda's Fragmentation

How do we conduct the wild giant panda census in dense forests and tall mountains? Why is the giant panda, which has survived for millions of years, fighting for its survival today?

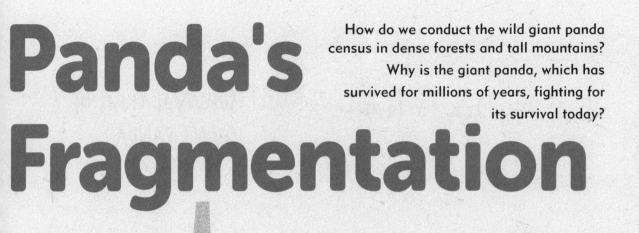

大熊猫的生存危机

"竹子开花啰喂,咪咪躺在妈妈的怀里数星星。星星呀星星真美丽,明天的早餐在哪里……" 20世纪80年代,很多小朋友都会唱这首歌。《熊猫咪咪》唱出了一个让人伤心的故事:竹子开花枯死,大熊猫"咪咪"没有吃的东西了……

1974年至1976年,国家林业部调查队在四川省多片开花后枯萎的竹林中,发现了138具大熊猫尸体。解剖发现,这些大熊猫的胃里没有食物残渣。1983年夏季,邛崃山系的箭竹大面积开花枯萎,500多只大熊猫再次面临饿死的危险。歌曲《熊猫咪咪》和相关新闻报道在国内外引发关注,掀起了"拯救大熊猫"的热潮。日本、美国、英国等很多国家都有人赶到中国为大熊猫的生存慷慨解囊,联合国粮农组织还启动了快速行动项目,无偿支援了88.7万美元。

作为大熊猫的主食,竹子每经历一定的周期,就会大面积开花。竹子开花,意味着这株竹子即将枯

SURVIVAL CRISIS OF GIANT PANDA

"Under a blossoming bamboo forest, the *Giant panda Mi Mi* counts the stars in her mother's arms. The stars are so beautiful, but where is the breakfast for tomorrow?..." It was a song sung by many children in the 1980s. *Giant Panda Mi Mi* told a sad story: as bamboo blossomed and withered, the giant pandas lost their food source...

From 1974 to 1976, the investigation team of the Ministry of Forestry found 138 dead pandas in withered bamboo forests across Sichuan. Autopsies revealed that the panda's stomachs were empty. In the summer of 1983, the *Fargesia spathacea* in the Qionglai Mountain blossomed and withered in large areas, threatening over 500 giant pandas with starvation. Thanks

竹子开花
Bamboo blossoming

to the song *Giant Panda Mi Mi* and news reports, an upsurge in support to save the giant pandas was incited across the world. The Chinese people were extremely concerned about this matter, people in Japan, the United States, the United Kingdom, and many other countries rushed to China to donate for the survival of giant pandas, and Food and Agriculture Organization of the United Nations (FAO) enacted a quick action project and provided USD 887,000 in support.

In fact, the giant panda's staple food bamboo blossoms and subsequently dies in a large area according to a certain cycle. When the bamboo blossoms, it means that it can no longer be consumed by giant pandas.

Will the giant pandas have no other food and suffer from starvation after the bamboo blossoms and dies? In fact, about 60 species of bamboo are edible to giant pandas around the world. The bamboo blossoming does not necessarily threaten the giant panda's survival, as each bamboo species has their own blossoming cycle. When one bamboo species blossoms, the giant panda can still migrate to other areas to search for alternative food sources.

开花的竹林
Blossomed bamboo forest

死,那么大熊猫也就不能食用了。

如果竹子都开花枯死了,大熊猫是不是就没有食物、将要挨饿了呢?其实,全世界可供大熊猫食用的竹子有60多种,竹子开花本身并不会威胁到大熊猫的生存,因为每种竹子都有自己不同的开花周期,其中一种竹子开花了,大熊猫可以迁徙去别处寻找其他种类的竹子食用。

但是，由于目前人口迅速增长，城镇快速扩张，森林被大规模砍伐，人们修建道路、进行旅游开发等，导致大熊猫的栖息环境遭到了严重破坏，它们的生存空间被大幅度压缩，原来的大片区域被分割成了一个一个的"小岛"。

如果某个"小岛"上的某种竹子开花了，生活在这个"小岛"上的大熊猫就很难迁徙到另外的"小

However, rapid population growth, urban development, large-scale deforestation, road construction, tourism development, and much more have severely destroyed and encroached upon giant panda habitats, fragmenting what were originally large areas of their living spaces into isolated "islands".

Habitat fragmentation makes it difficult for giant pandas living on one "island" to migrate to another to find alternative bamboo species when a certain bamboo species in the "island" blossoms, thus adding to the plight of food shortage.

不同种类竹子开的花
Blossoms from different kinds of bamboo

现在，大熊猫呈孤岛状分布于秦岭、岷山、邛崃山和大、小相岭及凉山6个山系的狭长地带。"孤岛"将大熊猫分割成了约33个孤立的小种群。

At present, isolated giant panda populations are distributed in corridors in the Qinling Mountains, Minshan Mountains, Qionglai Mountains, Daxiangling Mountains, Xiaoxiangling Mountains, and Liangshan Mountains. Habitat fragmentation has divided the giant panda population into about 33 isolated small populations.

岛"去寻找其他种类的竹子，从而陷入缺少食物的困境。而到了谈恋爱的季节，大熊猫也只能在自己所在的"小岛"寻找心仪的对象，长此以往会导致大熊猫近亲繁殖，使种群繁衍出现问题。

此外，偷猎、采笋等活动，对大熊猫的生存也造成了严重威胁，致使大熊猫的数量越来越少，生存境况越来越糟糕。

In addition, during estrus season, their options for mating are limited to the "island", which may lead to inbreeding and affect population quality if the problem remains.

In addition, factors like people poaching and harvesting bamboo shoots have also posed a serious threat to giant panda survivability, resulting in a declining panda population and worsening situation.

大熊猫"人口普查"

THE GIANT PANDA CENSUS

为了查清野外大熊猫到底有多少,2011年,第四次全国大熊猫"人口普查"工作开始了。

别看大熊猫平时慢腾腾的,其实它们在山林里非常机警,行动迅速。它们的嗅觉和听觉都很厉害,察觉到有人来后马上就会躲起来。要想在野外见到大熊猫那是要靠运气的。

所以,调查队员们只能在大熊猫栖息地追踪大熊

To find out how many giant pandas there are in the wild, the Fourth National Giant Panda Census was launched in 2011.

Do not be deceived by their usual slow movement, giant pandas are actually alert and swift in mountains and forests. They come equipped with keen senses of hearing and smell and can quickly hide once they sense anything incoming. You need to rely on luck to spot giant pandas in the wild. Therefore, the investigators can only

野外调查
Field survey

野外调查途中涉水过河
Trekking Across River during Field Survey

track clues of their activities including feces, paw prints, and bamboo stem fragments to infer giant panda population size.

After two years of arduous field surveys, the number of wild giant pandas was finally announced in 2015. The survey results revealed that the wild giant panda population in China totaled 1,864, increased by about 200 than a decade ago.

The 1,864 wild giant pandas have inhabited the Qinling Mountains, Minshan Mountains, Qionglai Mountains, Daxiangling Mountains, Xiaoxiangling Mountains, and Liangshan Mountains across Sichuan, Shanxi, and Gansu provinces and about 75% of them live in Sichuan Province.

猫的生活痕迹，通过找粪便、查脚印、看它们吃剩的竹子，来判断大熊猫的数量。

经过两年艰苦的野外调查，野生大熊猫的数量终于在2015年公布了。当年，全国野生大熊猫的数量为1864只，比10年前增加了200多只。

这1864只野生大熊猫生活在秦岭、岷山、邛崃山、大相岭、小相岭和凉山这6大山系中，横跨四川、陕西、甘肃三省，其中近75%生活在四川境内。

大熊猫小百科

全国野生大熊猫调查

大约每隔10年，我国就会进行一次野生大熊猫调查，每次调查耗时3到5年。通过野外调查，我们可以了解大熊猫的种群数量和栖息地范围。截至2020年，全国已经进行过4次野生大熊猫调查。

GIANT PANDAS TIDBITS

National Wild Giant Panda Population Survey

A wild giant panda population survey that will take three to five years is organized and conducted almost every decade. Field surveys are a way to learn about giant panda population quantities and habitat ranges. Before 2020, four nationwide surveys had been conducted.

调查次序 Survey order	年份 Year	数量（只）Number of Pandas
第一次调查 The First Survey	1974—1977	2459
第二次调查 The Second Survey	1985—1988	1114
第三次调查 The Third Survey	1999—2003	1596
第四次调查 The Fourth Survey	2011—2013	1864

NO.4 大熊猫的"孤岛"困境

保护区的红外相机拍摄到的野生大熊猫
Wild giant panda captured by infrared cameras in the reserve

野外调查进山途中
Into the Mountains to Conduct Field Survey

What's It Be a Giant Neighbor

NO.5 The Giant Panda You Don't Know

和"团子"做邻居是啥感觉

大熊猫有哪些动物朋友？
1.8 米的大高个儿，嗅觉、听力都不错，又很擅长爬树，这么厉害的大熊猫在野外有谁敢招惹？

Like to Panda's

What animals can live in harmony with giant pandas? Due to a height of up to 1.8 m, superior sense of smell and hearing, and great gift for climbing trees, giant pandas maintain an unrivaled position in the wild and who dares to provoke them?

大熊猫的朋友们

GIANT PANDA'S FRIENDS

    与大熊猫生活在同一片栖息地，在食物、栖息环境、水源地、隐蔽条件等资源利用方面和大熊猫互补，或者相互影响，这样的动物被叫作大熊猫伴生动物。

    大熊猫生活的地方风景可美了！那里有高山密林，还有清澈的流水，云雾缭绕，仿佛仙境一般。

    这些地方海拔1500～3500米，因为海拔比较高，所以气温适宜，常年都在20℃以下。

    在这样一个超级舒适的"生态小区"里，低调的大熊猫安安静静地吃着竹子，舒舒服服地睡着大觉，一副与世无争的样子。这里还居住着川金丝猴、羚牛、小熊猫、毛冠鹿、红腹锦鸡等珍稀动物。它们各自占有自己的生活空间，在食物选择和营养需求上互不影响，快乐、和谐地生活在一起。

Refers to an animal that shares the same habitat with the giant panda and has established a complementary relationship with or exerts mutual influence on it in respect to utilizing food, habitat, water sources, shelter, and other resources. These species called giant panda sympatric species.

The beautiful homeland of wild giant pandas is host to mountains shrouded in mist, thick forests, and clear water, like a fairy tale.

A high altitude of approximately 1,500 m to 3,500 m gives rise to an agreeable temperature that sits below 20 ℃.

Giant pandas live a harmonious life where they quietly enjoy bamboo and contentedly sleep in such an advantageous ecological environment, while always keeping a low profile. This area is also home to other rare animals, like the golden snub-nosed monkey, takin, red pandas, tufted deer, and golden pheasant definite with activity spaces, exerting no influence on each other's food and nutritional demands. All of the species coexist with one another in a happy and harmonious way.

> 小熊猫
>
> 英文名：Red Panda
>
> 拉丁学名：*Ailurus fulgens*
>
> 食物：竹笋、竹叶、植物的果与叶、小型动物等；
>
> 外形特点：全身红褐色，四肢黑褐色，脸颊有白色斑纹，尾巴有9~12个红白相间的环纹；
>
> 分布：主要分布于中国西南地区及周边国家海拔1800~4800米的温带森林；
>
> 寿命：10~12岁；
>
> 保护等级：IUCN濒危物种，中国国家二级保护动物。

Red Panda

English name: Red Panda

Scientific name: *Ailurus fulgens*

Diet: Bamboo shoots, bamboo leaves, fruits, plant leaves, small animals, etc.;

Appearance: Covered with reddish brown fur with dark brown limbs, cheek with white markings, and tail with 9-12 red and white rings;

Distribution: Primarily distributed in temperate forests at altitudes of 4800-1,800 m in southwest China and neighboring countries;

Lifespan: 10-12 years;

Protection class: Listed as an endangered species on the IUCN Red List and an animal under Class II state protection in China.

川金丝猴

英文名：Golden Snub-nosed Monkey

拉丁学名：*Rhinopithecus roxellana*

食物：松针、苔藓、竹笋、植物的嫩芽与野果等；

外形特点：毛发金黄，脸部为天蓝色，鼻孔朝天；

分布：主要分布于中国西南地区海拔3400米以下的竹林和针叶林；

寿命：10～25岁；

保护等级：IUCN濒危物种，中国国家一级保护动物。

Golden Snub-nosed Monkey

English name: Golden Snub-nosed Monkey
Scientific name: *Rhinopithecus roxellana*
Diet: Pine needles, moss, bamboo and plant shoots, and wild fruits, etc.;
Appearance: Golden fur, sky-blue face, and upturned nostrils;
Distribution: Primarily distributed in bamboo and coniferous forests at altitudes below 3,400 m in southwest China;
Lifespan: 10-25 years;
Protection class: Listed as an endangered species on the IUCN Red List and an animal under Class I state protection in China.

红腹锦鸡

英文名：Golden pheasant

拉丁学名：*Chrysolophus pictus*

食物：各种植物的花、果、叶、芽、种子，以及小型昆虫等；

外形特点：雄鸟羽色华丽，赤橙黄绿青蓝紫俱全；

分布：主要分布于中国秦岭山脉海拔500～2500米的阔叶林和灌丛地带；

寿命：15～20岁；

保护等级：IUCN低危物种，中国国家二级保护动物。

Golden Pheasant

English name: Golden Pheasant
Scientific name: *Chrysolophus Pictus*
Diet: Flowers, fruits, leaves, buds and seeds of various plants, small insects, etc.;
Appearance: The male boasts colorful lush plumage with different colors including red, orange, yellow, green, cyan, blue, and purple;
Distribution: Primarily distributed in broad-leaved forests and shrub zones at altitudes of 500-2,500 m in China's Qinling Mountains;
Lifespan: 15-20 years;
Protection class: Listed as a least concerning species on the IUCN Red List and an animal under Class II state protection in China.

羚牛

英文名：Takin

拉丁学名：*Budorcas taxicolor*

食物：竹笋、竹叶、各种植物的枝叶与树皮等；

外形特点：全身毛色金黄，角尖向内扭曲，体壮如牛，头小尾短似羚羊；

分布：主要分布于中国西南、西北地区及周边国家海拔2500米以上的高寒密林；

寿命：12～15岁；

保护等级：IUCN易危物种，中国国家一级保护动物。

Takin

English name: Takin

Scientific name: *Budoreas taxicolor*

Diet: Bamboo shoots, bamboo leaves, branches and leaves of various plants, bark, etc.;

Appearance: Golden yellow fur, inwardly twisted horns, a strong body, and a small head and short tail like an antelope;

Distribution: Primarily distributed in dense alpine forests at altitudes above 2,500 m in southwest China, northwest China, and neighboring countries;

Lifespan: 12-15 years;

Protection class: Listed as a vulnerable species on the IUCN Red List and an animal under Class I state protection in China.

THE THREATS

Apart from interesting and friendly animals, giant pandas also meet threats sometimes, like the Asian golden cat, dhole, yellow-throated marten, and leopard.

Sporadic oversights in security will offer these threats an opportunity to attack old and feeble giant pandas, as well as cubs. They obviously target the weak, but they are no match for young and strong giant pandas with ferocity inherited from their carnivorous ancestors.

"讨厌"的家伙们

在大熊猫的野外家园，有趣的动物朋友很多，但是和它们生活在一起的还有一些"讨厌"的家伙，比如金猫、豺、黄喉貂、豹。

这个"生态小区"的安保工作偶尔也有出现疏漏的时候，那些"讨厌"的家伙们就会趁机袭击年幼和年老体弱的大熊猫——这明摆着是在欺负弱者嘛！如果遇到年轻体壮的大熊猫，它们则根本不是对手，毕竟大熊猫仍然保留着食肉祖先的凶猛本性。

金猫

英文名：Asian Golden Cat

拉丁学名：*Catopuma temminckii*

食物：啮齿动物、鸟类、幼兔，麂和麝等小型鹿类；

外形特点：体毛多为棕红或金褐色，两眼内侧各有一条白纹，额部有带黑边的灰色纵纹，延伸至头后；

分布：主要分布于东南亚、中南半岛和中国的热带和亚热带湿润常绿阔叶林、混合常绿山地林和干燥落叶林；

寿命：10～20岁；

保护等级：IUCN近危物种，中国国家一级保护动物。

Asian Golden Cat

English name: Asian Golden Cat
Scientific name: *Catopuma temminckii*
Diet: Rodents, birds, young rabbits, muntjacs and musk deer, and other small animals in the Cervidae family;
Appearance: Mostly red or golden brown, with a white stripe beside each eye, and gray vertical stripes with black edges that extend from the forehead to the back of the head;
Distribution: Primarily distributed in humid evergreen broad-leaved forests, mixed evergreen mountain forests and dry deciduous forests of tropical and subtropical climates in Southeast Asia, the Indo-China Peninsula, and China;
Lifespan: 10-20 years;
Protection class: Listed as a nearly threatened species on the IUCN Red List and an animal under Class I state protection in China.

豺

英文名：Dhole

拉丁学名：*Cuon alpinus*

食物：主食鼠、兔等小型兽类以及马、鹿等有蹄类动物，偶尔也吃甘蔗、玉米等素食；

外形特点：头背部毛色棕褐，腹部及四肢内侧为浅白色，尾巴粗而蓬松；

分布：广泛分布于北亚、南亚及东南亚的大陆地区；

寿命：15～16岁；

保护等级：IUCN濒危物种，中国国家一级保护动物。

Dhole

English name: Dhole
Scientific name: *Cuon alpinus*
Diet: Mostly small animals, such as rats and rabbits, and ungulates, such as horses and deer, and occasionally vegetables, such as sugar cane and corn;
Appearance: Brown fur on the head and back, pale white fur on the abdomen and inside the limbs, and a thick and fluffy tail;
Distribution: Widely distributed in North Asia, South Asia, and other regions of Southeast Asia;
Lifespan: 15-16 years;
Protection class: Listed as an endangered species on the IUCN Red List and an animal under Class I state protection in China.

黄喉貂

英文名：Yellow-throated Marten

拉丁学名：*Martes flavigula*

食物：鼠、獾、狸、鸟和鸟蛋、鱼，以及植物的果实等，还可合群捕杀麝、鹿等大型兽类；

外形特点：因前胸部具有明显的黄橙色喉斑而得名，头较为尖细，略呈三角形，耳短而圆，体形细长；

分布：主要分布于东亚和东南亚及俄罗斯外东北地区的林区；

寿命：14岁左右；

保护等级：IUCN无危物种，中国国家二级保护动物。

Yellow-throated Marten

English name: Yellow-throated Marten
Scientific name: *Martes flavigula*
Diet: Rats, badgers, raccoon dogs, birds and their eggs, fish, fruit, etc. They can hunt large animals, such as musk deer and deer, in packs;
Appearance: Marked by a rather pointed and slightly triangular head, short and round ears, and a slender body, it is named for the obvious yellow-orange patch on its throat and front chest;
Distribution: Primarily distributed in the forests of East Asia, Southeast Asia, and the Outer Northeastern Region in Russia;
Lifespan: About 14 years;
Protection class: Listed as a least concerning species on the IUCN Red List and an animal under Class II state protection in China.

豹

英文名：Leopard

拉丁学名：*Panthera pardus*

食物：有蹄类动物，也捕食猴、兔、鼠类、鸟类和鱼类，秋季还采食甜味浆果；

外形特点：被毛黄色，满布黑色环斑，头部斑点小而密，背部斑点密而大，斑点为圆形或椭圆形的梅花状图案；

分布：广泛分布于亚非，生活在森林、灌丛、湿地、荒漠等环境中；

寿命：14～19岁；

保护等级：IUCN易危物种，中国国家一级保护动物。

Leopard

English name: Leopard

Scientific name: *Panthera pardus*

Diet: In addition to ungulates, it also preys on monkeys, rabbits, rodents, birds, and fish, then feeds on sweet berries in the autumn;

Appearance: A yellow coat with dense black ring spots that are small on its head and large on the back. The spots are round or oval patterns that resemble plum blossoms;

Distribution: Forests, thickets, wetlands, deserts, and other environments across many parts of Asia and Africa;

Lifespan: 14-19 years;

Protection class: Listed as a vulnerable species on the IUCN Red List and an animal under Class I state protection in China.

The Giant Panda You Don't Know

NO.6

Do Giant Hibernate

大熊猫会冬眠吗

大熊猫会像其他熊一样冬眠吗?
大熊猫怕热还是怕冷呢?

Pandas

Do giant pandas hibernate like other bears? Are giant pandas intolerant of heat or cold?

大熊猫为啥不怕冷？

WHY GIANT PANDAS ARE NOT INTOLERANT OF THE COLD?

北风那个吹，雪花那个飘；雪花那个飘飘，冬天来到。

Winter comes when the north wind roars and snowflakes flutter about.

寒风凛冽，漫天飞雪，人们待在屋里哪儿都不想去，一些动物也选择通过冬眠来度过漫长的冬季。大熊猫却是个例外：它们不仅不怕严寒，不需要冬眠，而且会撒欢儿似的在风雪中玩耍，轻轻松松地出去觅食。

这就让人纳了闷儿了：大家都怕冷，为什么大熊猫不怕冷呢？

Bitterly cold wind and snowy weather keep people cooped up indoors. Some animals establish a routine of hibernation during the long winter, but it is not the case with giant pandas. They prefer to enjoy themselves in the wind and snow rather than hibernate. The cold weather is by no means a handicap when it comes to foraging.

It is strange that all animals are intolerant of the cold, but the giant panda is an exception. Why is that?

摄影：张志和
Photo by Zhang Zhihe

摄影：张志和 Photo by Zhang Zhihe

Reason 1 : The Distinctive Black and White Coat

It is attributed to their distinctive black and white coat. It is not only attractive but also warm enough to help giant pandas perfectly resist the cold. It looks soft and glossy but proves to be pretty thick, bushy, and rigid and is comprised of guard fur and under fur. According to microscopic fur findings, a thick inner medulla contributes to terrific heat preservation, while the outer layer that is enriched with fat shows a favorable moisture-proof functionality. Giant pandas are fully covered in this protective fur, so they can adapt to humidity and severe cold and remain unaffected when tumbling in the snow.

原因一：黑白限定款外套

这就得说说"团子"们亮眼的黑白限定款外套了。这件外套可是顶级的冬装，有风度更有温度，黑白相间的皮毛看上去柔软而光滑，实际上却非常厚密、硬直。大熊猫的被毛厚而粗，由针毛和绒毛组成。从毛的显微结构看，其内层髓质层比较厚，有良好的保温性能，外层富含油脂，具有良好的防潮作用。这样的皮毛把大熊猫完完全全地包裹并保护了起来，让它们适应了潮湿、严寒的自然环境，让它们可以在雪地上任意翻滚而安然无恙。

摄影：张志和
Photo by Zhang Zhihe

原因二：冬天充足的食物

当然啦，大熊猫不需要冬眠还有另外一个重要原因，那就是食物充足。中国可是响当当的"竹子王国"，不管是竹子种类、竹子生长面积、竹子产量都是世界第一。适合在热带和亚热带的高海拔、高纬度地区生长的竹子，我们都有。全世界70多属1200多种竹子，我国就生长着其中差不多一半的品种。竹子种类多，一年四季都有适合大熊猫采食的种类，它们当然就不需要冬眠啦！

Reason 2 : Plenty of Food in the Winter

The other important reason why hibernation is unnecessary for giant pandas is an abundance of food. The world's top bamboo varieties, areas, and yields in China crown the country with the title of Bamboo Kingdom. Plus, bamboo is widely dispersed in tropical, subtropical, high-altitude, and high-latitude areas of China. Nearly half of the over 1,200 varieties of bamboo across more than 70 genera on the earth may be found in China. A great variety of bamboo is available throughout the year, which spares the giants pandas from hibernating.

NO.6 Do Giant Pandas Hibernate　　NO.6 大熊猫会冬眠吗

啥也不怕，只怕酷暑

"冰块里藏着什么好吃的？"
"What's hidden in this ice cube?"

黑白条纹外套只有冬款。再冷，大熊猫都不怕；但是，炎热，它们就完全没办法应对了。

身为猛兽，大熊猫天不怕地不怕，就是超级惧怕酷暑。夏天是它们最难熬的季节。在野外，如果气温升高，它们会毫不犹豫地搬家，从低海拔地区迁移到高海拔地区，这样就会凉快一些；如果在动物园里，一般情况下，当气温高于25 ℃时，它们就会回到空调房里享受清凉。

SCARED OF NOTHING BUT THE HOT SUMMER

The black and white coat helps giant pandas resist severe cold but is of no use in hot weather.

As a famously savage beast, giant pandas are scared of nothing but the hot summer, which is the hardest time for them. Wild giant pandas will be bound to move from low altitudes to cool places at high altitudes when temperatures begin to increase; pandas in captivity will prefer to stay in air-conditioned rooms when temperatures exceed 25 ℃.

"好凉快！"
"So cool!"

大熊猫小百科

黑白限定款大熊猫外套

　　胖嘟嘟、萌萌哒的大熊猫以一身独特的黑白装扮圈粉无数。在动物时尚界，大熊猫这身黑白相间的皮毛很少有"撞衫"的时候。科学家研究发现，大熊猫的黑白装可不仅仅只是好看，更主要的"设计理念"是保障生存呢！

　　大熊猫皮毛的黑、白两色，在雪地或森林中能起到保护色的作用，帮助大熊猫躲避天敌。研究人员通过比较发现，大熊猫的脸、脖子、肚皮和臀部之所以是白色，是因为这能帮助它们在积雪覆盖的栖息地躲藏；而黑色的四肢，则能帮助它们将自己躲藏在阴影里。也有研究人员认为，黑白相间的毛色便于大熊猫在野外光影斑驳的森林中穿行自如。

GIANT PANDA TIDBITS

Distinctive Black and White Coat

Chubby and adorable giant pandas with a special black and white coat have won over a wealth of fans. As a unique symbol of giant pandas, this coat is more functional than aesthetic. Based on scientific research, it is primarily used to guarantee survival.

The black and white color protects giant pandas from being identified by predators in the snow or forests. Based on comparative research discoveries, their white faces, necks, bellies, and hips may help them hide in snow-covered habitats while their black limbs are beneficial for hiding among the shadows. Some researchers have given different theories that the black and white color has evolved to help giant pandas freely pass through forests under dappled light.

NO.7 — The Giant Panda You Don't Know

Distinction Giant and Red

大、小熊猫可别傻傻分不清

你是否以为，小熊猫和大熊猫的中文名字只有一字之差，所以它俩是一家人？
Panda（熊猫）最早是哪种动物的英文名字？

Between Panda Panda

Do you think giant pandas and red pandas belong to the same genus since their Chinese names are only different by one character? Which animal was initially named the panda?

大熊猫和小熊猫是两种动物

GIANT PANDA AND RED PANDA ARE TWO KINDS OF ANIMALS

小熊猫就是小小只的大熊猫宝宝吗?

提到小熊猫,你脑海中浮现出的是什么模样呢?是黑白相间的大熊猫宝宝呢,还是红棕色的小熊猫?其实,大熊猫和小熊猫只是名字有点儿像而已,它们是压根儿不同的两种动物,既没有兄弟姐妹关系,也没有父母子女关系。究竟小熊猫和大熊猫有着怎样的区别与联系呢?一两句话可说不清楚,还是听听它们是怎么介绍自己的吧。

Does the red panda refer to a small giant panda cub?

What images come to mind when you think of the red panda? A black and white giant panda cub or a reddish-brown red panda? In fact, the red panda and giant panda have similar Chinese names yet are notably different. They are neither siblings nor parent and child. What are the differences and relationship between the red panda and giant panda? It cannot be specified in a few words. It is advisable to refer to their profiles.

一看户籍、身份
The First is Different Distributions

我们主要生活在中国的四川、陕西、甘肃三省，根据第四次大熊猫"人口普查"的结果，目前我们野外的同类数量为1864只，在《世界自然保护联盟濒危物种红色名录》中被列为易危物种。

We're endemic to Sichuan Province, Shanxi Province, and Gansu Province. As the wild population totals 1,864 according to the fourth giant panda census, we are a vulnerable species under the IUCN Red List.

我们主要生活在中国、印度、老挝、缅甸、尼泊尔、不丹等亚洲国家。在中国主要生活在西藏（喜马拉雅山南坡）、云南、四川等地。目前，我们在全球的数量已经不足1万只了，在《世界自然保护联盟濒危物种红色名录》中被列为濒危物种。

We are primarily distributed in China, India, Laos, Myanmar, Nepal, Bhutan, and other Asian countries. Most of us in China may be found in Tibet (southern slopes of the Himalayan Mountains), Yunnan, and Sichuan. A total population of less than 10,000 has made us listed as an endangered species under the IUCN Red List.

大熊猫的分布区域
The distribution of giant panda

小熊猫的分布区域
The distribution of red panda

二看生活习性
The Second is Different Living Habits

我们99%的食物是竹子,在野外偶尔还吃一些动物的尸体或其他植物。

Bamboo accounts for 99% of our diet, and we may occasionally ingest some dead animals or other plants in the wild.

我们主要吃竹笋、竹叶和嫩枝,在野外还要吃野果、树叶、苔藓、鸟蛋,会捕食小鸟和其他小动物。

We live on bamboo shoots, bamboo leaves, and twigs. When we live in the wild, our diet may be supplemented by wild fruit, tree leaves, mosses, birds, bird eggs, and other tiny animals.

我们4~6岁成年,这时候身长能达到160~180厘米,体重能达到80~150千克。在野外我们能活到18~20岁;而在动物园里,因为被很好地照顾,一般能活到25岁左右,甚至更高寿。

Adulthood begins from 4-6 years old, when we are as long as 160-180 cm and weigh near 80-150 kg. Compared to those in the wild with a lifespan of 18-20 years, we may live up to nearly 25 years old in zoos, maybe even longer thanks to good care.

我们1.5~2岁成年,这时候的身长有40~63厘米,体重5千克左右。我们一般能活到12岁半左右。

We enter adulthood at the age of 1.5 or 2 when we are as long as 40-63 cm and weigh near 5 kg, and our general lifespan is approximately 12.5 years.

三看成长变化
The Third is Different Changes in Growth

我们刚出生时皮肤是粉色的,长着稀疏的白毛。我们长大后,皮毛是黑白相间的。我们的尾巴是白色的、扁扁的,很短,像一把刷子。

We are pink with sparse white fur at birth. We are known for black and white fur, as well as a white, flat, and short tail like a brush when we grow up.

我们刚出生时,浑身是浅灰色的,毛茸茸的。当我们长大后,皮毛会变成红褐色。我们还有一条漂亮的长尾巴,尾巴上有9~12个红白相间的环纹。

We are light gray and furry at birth. We will have reddish-brown fur and a beautiful long tail with nine to twelve red and white circular rings when we grow up.

大熊猫和小熊猫的成长

The Growth of Giant Panda and Red Panda

了解了吧，我们是两种不同的动物！

If you know more, you will find we are a far cry from each other.

千万不要再把我们弄混了！

Don't confuse us!

嗯，讲得挺清楚的。只是……只是……还有一个问题：那些不大不小的大熊猫也不适合喊大熊猫宝宝，难道得叫它们"小大熊猫"吗？

It is quite clear. However, there is still a small question. For those medium-sized giant pandas that are not cubs anymore, can I call them a "small giant panda"?

虽然我就快2岁了，但你要叫我"宝宝"，我也没意见。

"Even though I'm almost 2 years old, I'm fine with you calling me baby."

NO.7 Distinction Between Giant Pandas and Red Pandas   **NO.7 大、小熊猫可别傻傻分不清楚**

FUN FACTS

趣闻一串串

小熊猫：抗议！我的名字被抢了！

1825年，法国动物学家弗列德利克·居维叶首次描述并命名了小熊猫，并且将其誉为自己见过的最美丽的动物。那个时候，小熊猫的英文名字就是Panda（熊猫）。

44年后的1869年，科学家发现了一种个头更大的熊猫，这种熊猫在身体结构上与小熊猫相似。为了区别这两种动物，科学家就把1825年发现的熊猫改称为小熊猫（Lesser panda）或红熊猫（Red panda），而把1869年发现的熊猫称为大熊猫（Giant panda）了。

Red Panda: I Protest that My Name Panda is Given to Others!

In 1825, Frédéric Cuvier, a French zoologist pioneered describing this species and gave it a name, praising it as the most beautiful animal he had ever seen. From then on, the species was dubbed panda in English.

44 years later, i.e. in 1869, scientists discovered another larger panda, which bears resemblance to the red panda internally. In a bid to differentiate them, the name of the panda discovered in 1825 was changed to lesser panda or red panda, while the panda identified in 1869 was dubbed the giant panda.

NO.7 Distinction Between Giant Pandas and Red Pandas　　NO.7 大、小熊猫可别傻傻分不清楚

NO.8 The Giant Panda You Don't Know

Unique

特殊食谱

春夏秋冬4个季节,"吃货"大熊猫吃竹子有哪些不同的讲究呢?
成都大熊猫繁育研究基地的熊猫窝窝头是什么味道的呀?

Diet

What are the giant panda's requirements for bamboo in different seasons? What does the panda cake of the Chengdu Research Base of Giant Panda Breeding tastes like?

大熊猫的主食——竹子

GIANT PANDAS PRIMARILY FEED ON BAMBOO

"西边来了一个怪,戴着墨镜好气派。拿着竹笛不去吹,张着嘴巴吃起来。"

地球人都知道,大熊猫的主食是竹子。竹子分布广泛,生长迅速,一年四季都能获得,因而受到大熊猫青睐。

"A monster from the western region wears stylish sunglasses. It doesn't play but eats the bamboo flute."

As is well known, giant pandas primarily feed on bamboo, which is widely distributed and available throughout the year.

NO.8 Unique Diet　　　　　　　　　NO.8 特殊食谱

白夹竹
Phyllostachys bissetii

箬竹
Indocalamus longiauritus

拐棍竹
Fargesia robusta

Bamboo is distributed all over the world with over 1,200 varieties across more than 70 genera, chiefly growing in tropical and subtropical areas, while a small number of them may be found in temperate and frigid areas.

China, as one of areas with the widest bamboo distribution, is home to over 400 varieties across nearly 40 genera, while over 80 varieties across 14 genera can be found in Sichuan. However, not all bamboo is palatable to giant pandas. They are only fond of over 20 varieties, including their staple food with the largest areas, like *Fargesia spathacea*, *Fargesia robusta*, *Bashania fangiana*, and *Chimonobambusa szechuanensis*.

全世界竹类植物有70多属，1200多种，主要分布在热带和亚热带地区，少数分布在温带和寒带地区。

中国是竹类分布最丰富的地区之一，约有40属400多种竹类，四川就有14属80多种。当然，并不是所有的竹子大熊猫都吃，它们喜欢吃的竹子只有20多种，其中箭竹、拐棍竹、冷箭竹、八月竹等分布面积最大，是大熊猫的主要食物来源。

冷箭竹
Bashania fangiana

巴山木竹
Bashania fargesii

苦竹
Pleioblastus amarus

动物界的"吃货"

说大熊猫是动物界的"吃货"是有道理的，这家伙既爱吃也会吃：不仅对吃的竹子种类有所选择，随着季节的变换，它们所采食的竹子部位也有所不同。

春季、夏季，大熊猫喜欢吃竹笋；秋季喜欢吃竹子当年新生的枝叶；冬季，则主要以竹竿为食，而且要弃掉竹梢。它们除了吃竹类植物，偶尔还吃一些其他植物的枝叶、果实，以及一些动物的尸体等。所以，在野外，它们能对食物进行充分选择，能够满足自身的营养需要。

摄影：张志和
Photo by Zhang Zhihe

A FUSSY EATER

Giant pandas have earned the title of a fussy eater because they are interested in and skilled at eating. They are particular about bamboo varieties, and the bamboo part they ingest varies across different seasons as well.

They will choose bamboo shoots in the spring and summer, young branches and leaves in the autumn, and bamboo stems in the winter, discarding the tops. In addition to staple bamboo, they occasionally ingest other plants, fruits, or dead animals; therefore, they boast a wide range of food options to gain adequate nutrition.

NO.8 Unique Diet

WHAT DO GIANT PANDAS EAT IN THE ZOO?

Giant pandas in captivity are provided with limited varieties of bamboo. Although bamboo still accounts for 99% of their diet, it results in inadequate food options and nutrition. This problem has been satisfactorily resolved by adding or supplementing with concentrated feed, vitamins, and microelements.

If you come to the Chengdu Research Base of Giant Panda Breeding, you will see professional nutritionists preparing panda cakes in the panda kitchen. It is also possible to catch a glimpse of a keeper feeding giant pandas apples. Panda milk, offered to panda cubs under 1.5 years old, is formulated by mixing milk and some microelements based on each cub's weight and physical conditions and is really popular. These efforts are a valid way to maintain giant pandas' health with balanced nutrition.

NO.8 特殊食谱

大熊猫在动物园里吃些啥？

生活在动物园里的大熊猫，虽然99%的食物仍然是竹子，但是人们能够提供给它们的竹子种类有限，无法完全满足它们的营养需求和对食物进行自由选择的需要，所以需要为它们补充一些精饲料、维生素和微量元素。

如果有机会来到成都大熊猫繁育研究基地，你就有机会在大熊猫厨房看到营养师给大熊猫们制作"窝窝头"，也有机会看到饲养员给大熊猫喂苹果。盆盆奶是为1岁半以内的大熊猫宝宝配制的，是宝宝们最喜爱的食物。饲养员将奶与一些微量元素混合，根据每个宝宝的体重和身体状况进行配制，这是为了让宝宝们营养均衡，让它们身体更健康。

圈养大熊猫的辅食和食物中添加的微量元素
Complementary food and trace elements added to the diet of captive giant pandas

大熊猫小百科

讲究的熊猫窝窝头

成都大熊猫繁育研究基地为满足大熊猫的营养需求制作的窝窝头,是将玉米、大米、燕麦、大豆、维生素等和一些微量元素混合后,制作成各种样式的窝窝头,再用蒸箱蒸制而成的。大熊猫营养师会根据每只大熊猫的不同需求来确定窝窝头的供应量。这窝窝头吃起来虽然没有咸味儿,但充满了谷物的醇香,还带着粗粮的嚼劲儿。看起来很好吃的样子,想尝尝不?

制作熊猫窝窝头

原料:玉米、大米、燕麦、大豆、小麦、植物油、微量元素和维生素。

Ingredients: corn, soybean, rice, oats, wheat, vegetable oil, minerals, and vitamins.

NO.8 Unique Diet　　　　NO.8 特殊食谱

GIANT PANDA TIDBITS

Panda Cake and Its Complex Recipe

The panda cake at the Chengdu Research Base of Giant Panda Breeding is a mix of corn, rice, oats, soybeans, vitamins, and some microelements, which make it nutrition-rich, and then steamed in a steam box. The cake supply amount depends on each panda's unique needs based on decisions from professional nutritionists. In spite of there being no salt, the cake is appealing due to its strong flavors and chewy cereals. Does the panda cake look delicious? Do you want to try it?

图片提供：pandapia
Provided by pandapia

熊猫窝窝头的制作方法：

1.清洗：原料需要经过认真清洗；

2.烘烤：将原料放到75℃~140℃的烘烤箱中烘干；

3.混合：磨粉后按照配方比例称量混合均匀；

4.塑形：用模具加工成形；

5.蒸煮：高温蒸煮，完成制作。

Method of making steamed panda cake:
1. Carefully clean raw materials;
2. Bake the raw materials in a 75 ℃ −140 ℃ oven to remove moisture;
3. Weigh and mix the ingredients according to the formula ratio after grinding;
4. Process and shape the ingredients in molds;
5. Steam at a high temperature to finish making the steamed panda cake.

FUN FACTS

趣闻一串串

大熊猫爱偷吃铜铁？

大熊猫也被称作"食铁兽"，这名字是怎么来的呢？

春秋战国时期（2700年以前）的《山海经》记叙，大熊猫很像熊，毛色黑白，产于邛崃山严道县（今四川荥经县），并说它食铜铁，故称"食铁兽"。

关于"食铁兽"这一说法，据推测，应该是当时野生大熊猫经常跑到山下居民家中偷吃锅里的食物，把锅啃变形或啃坏了，所以被误认为能吃铜铁，后来便被讹传成了"能舐食铜铁，须臾便数十斤"。

Do Giant Pandas Like Eating Copper and Iron?

Where does the panda's other name "iron-eater" come from?

As recorded in the *Classic of Mountains and Rivers* written in the Spring and Autumn Period and the Warring States Period (2,700 years ago), the giant panda, with its bear-like and furry black & white body originated from Yandao County (present-day Yingjing County, Sichuan) in Qionglai Mountains and ate copper and iron. It is the accepted reason for why it is dubbed the "iron-eater".

It is speculated that the wild giant pandas might frequent residences at the foot of mountains and bit iron pots and pans to get food remnants inside them. When residents saw the deformed iron pots and pans, they thought that giant pandas were fond of eating iron, and it evolved into a myth that "giant pandas gain weight after ingesting copper and iron."

The Giant Panda You Don't Know

NO.9

Magical ?

神奇的伪拇指

看过了那么多大熊猫后,你知道它们有几根"手指"吗?
大熊猫版吃竹子教程,你听说过吗?

Pseudo-thumb

You have seen so many giant pandas.
Do you know how many fingers they have?
Have you heard of how giant pandas
eat bamboo?

大熊猫有六根"手指"

大熊猫能够非常灵活地拿取竹子,这让人们误以为,它们的前掌一定有着类似人类手掌的结构。实际上,大熊猫有六根"手指"。

严格地说,大熊猫的第六根"手指"并不是真正的"手指",因为它没有指甲,只是一个能够活动的凸起,被称为伪拇指。

"看到我的伪拇指了吗?"
"Did you see my pseudo-thumb?"

摄影:张志和
Photo by Zhang Zhihe

大熊猫的第六根"手指"
The giant panda's pseudo-thumb

GIANT PANDAS HAVE SIX FINGERS

Ordinarily, giant pandas can flexibly grab bamboo, which has given people the false impression that they have human-like fingers. In fact, they have six fingers.

Strictly speaking, the sixth finger can not really be counted as a finger because it is just a bulge that is able to move and does not have a fingernail, and is called "pseudo-thumb".

THE SIXTH FINGER PLAYS A VERY LARGE ROLE

It appears unimportant but actually plays a very large role as the giant panda's secret weapon!

The pseudo-thumb was generated due to wrist sesamoid becoming specialized and allows giant pandas with a unique function to fully grasp objects, which allows them to stand out from other bears.

The rare full-grasping function is exclusive to koalas, North American opossums, a majority of primates, giant pandas, and red pandas, and only the latter two species are endowed with a sixth finger which enables them to fully grasp objects with the other five fingers.

"第六指"有大作用

你可不要小看这根伪拇指，它是大熊猫抓握的秘密武器哟，它的作用非常大！

伪拇指是腕部籽骨特化生成的，它让大熊猫拥有了其他熊类不具有的对握功能，可以更好地进行抓握，灵活地采食竹子。

这种对握本领并不常见，除了考拉、北美负鼠和大部分灵长类动物，只有吃竹子的大、小熊猫才有。这其中，只有大熊猫和小熊猫特化出了"第六指"去和其他"五指"对握。

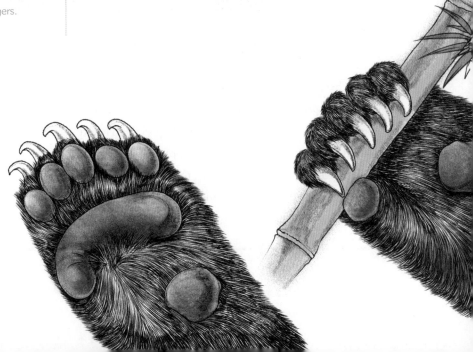

大熊猫小百科

大熊猫版免费品竹教程

对我们大熊猫来说,小小一根竹子吃起来可是有大学问的。

我们吃竹竿时,主要吃竹竿的中间部位。如果竹竿比较粗,就像人类啃甘蔗一样,用前掌握着竹竿,先把皮剥掉,然后再吃。试试看,这样口感是不是很好呢?

如果吃的是竹叶,要先聚集足够多的竹叶放在嘴边,然后再大口大口地吃。这样是不是很节约时间呢?

特别说明:此教程仅适合有伪拇指者。正是因为伪拇指的重要作用,我们才能灵巧、方便地花式进食。

GIANT PANDA TIDBITS

Free Tutorial on Giant Pandas Eating Bambo

There are many things worth mentioning about giant panda when eating bamboo.

Giant pandas mostly choose the middle part of bamboo stems. In case a bamboo stem has a larger diameter, giant pandas tend to peel them before eating to make them tastier.

Regarding bamboo leaves, giant pandas will gather a bunch of leaves beside their mouths before gobbling them down to save time.

Remark: The tutorial is merely intended for eaters with a pseudo-thumb. The pseudo-thumb is of great use and allows us to eat conveniently and freely.

NO.9 Magical Pseudo-Thumb NO.9 神奇的伪拇指

The Giant Panda You Don't Know

NO.10

Giant Teeth

牙齿那点儿事

硬竹子轻松嚼，大熊猫到底长了一口怎样特别的牙齿？

小孩子五六岁换牙，大熊猫也会换牙吗？

Panda's

It is amazing that giant pandas can easily chew through hard bamboo! What is special about giant pandas' teeth? Our baby teeth fall out at the age of 5 or 6. Does it work for giant pandas?

强大的咬合力

一个人悠悠闲闲地拿根竹子塞进嘴里,咔嚓一声咬下去……哎哟,想想都牙痛!大熊猫99%的食物是竹子,竹子不仅坚硬而且富含植物纤维,可这又硬又不好嚼的竹子大熊猫咬起来一点儿都不费劲,这是为什么呢?

让我们来一探究竟吧!

POWERFUL BITE FORCE

Seeing the giant panda biting bamboo and upon hearing the crunches, we cannot help but worry about their teeth! Bamboo enriched with plant fiber accounts for 99% of the giant panda diet. Why can giant pandas easily chew through the hard and chewy bamboo?

Let's uncover the mystery!

图片提供:pandapia
Provided by pandapia

Let's carefully compare giant panda and black bear skulls.

The giant panda's skull features a well-developed and broad zygomatic arch, which is conducive to having strong masseter attached when compared to that of the black bear; the giant panda has broad and strong molars and many tiny tooth tips on the surface that facilitate giant pandas chewing cellulose-rich bamboos.

Apart from functional adaptations, the giant panda's skull has evolved to allow giant pandas to easily gnaw through bamboo stems and chew through bamboo in a short time.

下图是大熊猫和黑熊的头骨，我们来比较一下。

大熊猫的头骨相比黑熊的头骨，颧弓更加发达也更加宽大，有利于强大的咬肌附着。此外，大熊猫的臼齿宽大粗壮，臼齿表面有许多细小的齿尖，有利于咀嚼高纤维的竹子。

大熊猫的头骨特征显示出对竹类食物的功能性适应，所以它们能轻松地咬断竹竿，快速地咀嚼竹子。

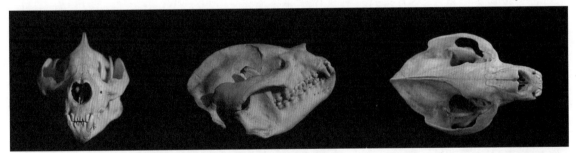

大熊猫的头骨
Giant panda skull

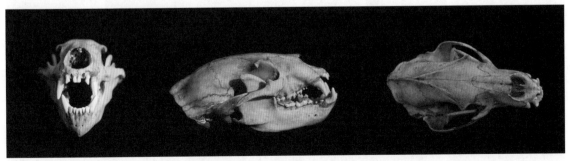

黑熊的头骨
Black bear skull

大熊猫有多少颗牙？

HOW MANY TEETH DO THEY HAVE?

大熊猫差不多在出生3个月后开始长牙，不过这时候它们还是完全以母乳为食，因为竹子太硬了，大熊猫宝宝是嚼不动的。

当它们5个月大时，乳牙全部长齐了，有24颗呢！人类的乳牙也就20颗，大熊猫宝宝的乳牙比人类还多了4颗。

过了没多久，也就8个月左右，大熊猫就开始换牙了，所以它们的乳牙长不了多久就会换掉。

Giant pandas begin teething when they are almost 3 months old, but they still rely on breast milk because bamboo is still too hard for them to chew.

When they are 5 months old, all 24 of their baby teeth have come in, 4 more than humans have.

When giant pandas are nearly 8 months old, the baby teeth fall out, a short time after they come in.

When they are about 1.5 years old, they have almost complete permanent

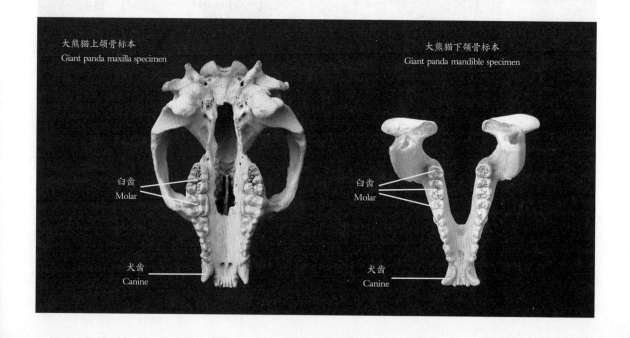

大熊猫上颌骨标本
Giant panda maxilla specimen

大熊猫下颌骨标本
Giant panda mandible specimen

白齿 Molar

犬齿 Canine

teeth, but different giant pandas may not have the same teeth in quantity. In certain cases, pandas have 40 or 42 teeth, depending on whether they have the last molars (i.e. molar teeth).

Good teeth are indispensable for a good appetite. In the course of wild animal conservation, the teeth, as a significant role player in the digestion process, shall be given much emphasis.

Giant panda's long-term consumption of hard bamboo makes teeth preservation and oral care particularly important. Oral disease examinations and healthcare given by dentists to giant pandas in zoos rescheduled on a regular basis.

到1岁半左右，它们的恒牙基本长齐了。悄悄告诉你们，每只大熊猫牙齿的数量不一定相同呢。有些大熊猫长不出左右两侧最后一颗白齿（大牙），这时候它们的牙齿是40颗；如果能长出左右两侧最后一颗白齿，那它们的牙齿就有42颗。

牙好，胃口就好，吃嘛嘛香。牙齿是动物重要的消化器官之一，好的牙口是动物健康的重要保证。

对长期食用硬竹子的大熊猫来说，牙齿健康和口腔护理尤为重要。所以，在动物园里，牙医会定期为它们进行口腔疾病检查和口腔保健。

大熊猫小百科

改吃素了犬齿仍然在

大熊猫的祖先"始熊猫"已经从肉食动物演化为杂食动物,那个时候它们的犬齿还发挥着重要的作用,竹子还不是它们的主要食物。后来,随着环境的变化,大熊猫祖先们的食物发生了巨大的变化。

虽然现在的大熊猫以竹子为主食,主要靠宽大的臼齿来咀嚼和磨碎竹子,但是它们依然保留着祖先们用于食肉的犬齿,只是因为很少使用,它们的犬齿已经钝化,不像虎、豹等肉食动物的犬齿那么锋利。

GIANT PANDA TIDBITS

The Canines Remain Although Pandas are Vegetarian Now

The giant panda's ancestor the *Ailurarctos lufengensis* had evolved from its ancestors the creodonts to omnivores, and canines played an important role at that time when bamboo had not become their staple food. The food of the giant panda's ancestors was subject to great changes due to varied environments.

Although giant pandas now primarily live on bamboo, and chew with broad molars, the ancestral canines that were used to chew meat remain. However, rare use has made the canines dull and not as sharp as those of tigers, leopards, and other carnivores.

NO.10 牙齿那点儿事
NO.10 Giant Panda's Teeth

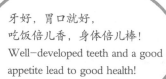

牙好，胃口就好，
吃饭倍儿香，身体倍儿棒！
Well-developed teeth and a good appetite lead to good health!

The Carnivora

NO.11 The Giant Panda You Don't Know

吃竹子的食肉目动物

大熊猫以竹子为主食,可是为什么在动物分类上属于食肉目动物呢?

在动物分类上,大熊猫到底属于哪一科?

Bamboo-Eating

Why is the giant panda that primarily feeds on bamboo classified as a carnivora? In the animal classification, what category does the giant panda belong to?

关于大熊猫食性的争论

DEBATE ABOUT THE GIANT PANDA'S FEEDING HABITS

科学课上，同学们对大熊猫是"植食动物"还是"肉食动物"展开了激烈讨论。

Whether the giant panda is a herbivore or a carnivore is fiercely debated in science class.

> 大熊猫主要吃的是竹子，从食性上看当然应该归入植食动物之列。

"The giant panda primarily feeds on bamboo, so it should be classified as a herbivore."

> 不对不对，动物的分类才没有这么简单呢。

"It is not correct. Animal classification is far from so simple."

NO.11 吃竹子的食肉目动物

Herbivore ?
Carnivore ?

我知道！我知道！早期人们将食性作为动物分类的重要参考指标，但是仅凭食性并不能对动物进行准确分类。

"I know. The feeding habit is designated as a significant reference index in the initial animal classification but cannot be accepted as a correct basis."

实际上，从消化道解剖结构、生理特点以及物种进化的角度来看，大熊猫是地地道道的食肉目动物。

"It is a true carnivora in terms of the anatomical structure of its digestive tract, physiological characteristics, and evolution."

改变食性以适应环境

同学们讨论过后，老师公布了答案：

"大熊猫的祖先是名副其实的肉食动物。它们不仅有尖锐发达的犬齿、较短的肠道，还具有食肉动物的消化生理特点。大熊猫现在仍保留着祖先的这些特征，只是由于生存环境发生了改变，为适应环境，它们最终选择了以低营养、低消化率的竹子为主食的生活。

"总而言之：从消化道特征来看，大熊猫不是植食动物；从食性来看，大熊猫也不是肉食动物。科学地说，大熊猫属于食肉目动物，是由有胎盘的肉食性哺乳动物经数百万年进化而来的。"

经过老师的讲解，同学们明白了：原来，不能只看动物吃的是植物还是肉类就简单地把它们划分为植食动物或肉食动物。

CHANGE THE FEEDING HABITS TO ADAPT THE CHANGED ENVIRONMENT

Following the students' discussion to its end, the teacher gives the answer:

"The giant panda's ancestor is a true carnivore, which boasted sharp and well-developed canine teeth, a short intestinal tract, and digestive characteristics like other carnivores. Giant pandas still possess characteristics from their ancestors, but to adapt to a greatly changed environment, they ultimately live on bamboo with low nutritional value and digestibility coefficients.

In conclusion, the giant panda is neither considered a herbivore from the perspective of its digestive tract nor a carnivore when judging from its diet. Scientifically speaking, the giant panda belongs to Order Carnivora that evolved over millions of years from placental carnivorous mammal."

Thanks to the teacher's explanation, the students ultimately understood that animals are by no means classified only based on whether they feed on plants or meat.

图片提供：pandapia
Provided by pandapia

FUN FACTS
趣闻一串串

你说说，大熊猫到底咋分类？

关于大熊猫的分类，科学家们一直争论不休。争论的原因在于分类依据不同。

一些科学家根据大熊猫特有的伪拇指结构，认为大熊猫与小熊猫相似，小熊猫属于浣熊科，因此大熊猫也应该属于浣熊科；一些科学家根据大熊猫的自然历史、生理特性等，认为大熊猫应该独立为一科，即大熊猫科；现在，更多科学家认为大熊猫的许多特征与熊相似，应该归于熊科。

In Your View, What Category does the Giant Panda Belong To?

The giant panda classification has been controversial for different scientists that specialize in various classification bases.

Some hold that the giant panda bears a close resemblance to the red panda based on the giant panda's unique pseudo-thumb structure, and the giant panda should be classified as a common raccoon, like the red panda. Others believe that the giant panda should belong to an independent family, i.e. the giant panda family in accordance to its natural history and physiological characteristics. Now, many experts support the giant panda belonging to the Ursidae because it exhibits a multitude of bear-like characteristics.

Spectacled Bear
眼镜熊

眼镜熊亚科
Tremarctinae

熊科
Ursidae

Giant Panda
大熊猫

大熊猫亚科
Ailuropodinae

NO.11 The Bamboo-Eating Carnivora NO.11 吃竹子的食肉目动物

American Black Bear
美洲黑熊

Asiatic Black Bear
亚洲黑熊

Polar Bear
北极熊

熊亚科
Ursinae

Brown Bear
棕熊

Sun Bear
马来熊

Sloth Bear
懒熊

NO.12 The Giant Panda You Don't Know

The Behind "Lazy Giant

我"懒"我有理

悠然自得，吃了睡，睡醒吃。
大熊猫为什么这么懒？
大熊猫一天中花在睡和吃上的时间有多长？

Reasons the Panda"

Giant pandas lead a rather leisurely life, which is composed of meals and sleep. Why are they so lazy? How much time do giant pandas spend on sleep and ingestion for one day?

"懒"是一种生存智慧

LAZINESS IS A LIFE WISDOM

动物园里,大熊猫小满睡得正香。突然,它被游客的声音吵醒了。它听到很多游客一边喊它起床,一边说它太懒了。小满并不在意大家对自己的看法,因为它知道,自己的"懒"是有原因的。让我们来听它说说吧!

When the giant panda Xiao Man was enjoying a sound sleep, he was awoken by some noise. Many visitors called for him to wake up and ridiculed his laziness. However, he did not really mind for there must be a certain reason to vindicate him.

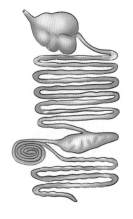

植食动物的肠道非常长，盲肠发达，如鹿的肠道大约是其体长的 25 倍。

Herbivores feature quite a long intestine and well-developed cecum. For example, the deer's intestine is about 25 times its body length.

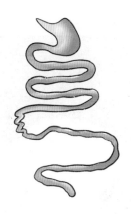

大熊猫的肠道直且短，没有盲肠，肠道大约是其体长的 4 倍。

The giant panda has a straight and short intestinal tract, which is about 4 times its body length, and lacks cecum.

"We are classified in the Order Carnivora under the Class Mammalia according to animal taxonomy, so we feature the structure of a carnivore's alimentary canal. Compared with many herbivores, our intestines are straight and short, so food can only remain in our alimentary canal briefly for 8-9 hours before being discharged in form of feces.

"我们在动物分类上属于哺乳纲食肉目动物，所以我们的消化系统当然也具有肉食动物消化系统的特点。与植食动物相比，我们的肠道不仅直而且短，所以食物仅在消化道内停留8～9小时就会变成粪便排出体外。

"我们既没有大容量的胃,也没有盲肠,这就不能为食物的消化发酵提供足够大的空间;同时,我们的消化道里还缺乏能消化纤维素的微生物。大多数食草动物能消化吸收食物的80%,而我们对食物的消化吸收率不足17%。

"吸收的营养少,还得维持庞大身体的能量需求,所以我们练就了一种本领:通过大量采食和让食物快速消化,从食物中最大限度地获取营养和能量,同时少动多休息,以有效地减少能量消耗。

"We have neither a large stomach nor a cecum, which prevents food digestion and fermentation. In addition, we lack microbes that digest cellulose. Our digestion and absorption ratio of food is less than 17% compared to 80% for most herbivores.

"To maintain normal functionality in our huge bodies with low nutrient absorption, we have developed a skill to get maximum nutrients and energy by large intake and fast passage of food. In addition, we barely move after eating to effectively save our energy.

NO.12 我"懒"我有理

"我们每天用于休息的时间几乎占了全天时间的一半,所以你平时会看到我们大部分时间都在睡觉,给人的印象是我们总是懒洋洋的。其实,这是我们的生存智慧,是为了保存实力。够聪明吧!"

"We spend nearly half our day sleeping, leaving people the impression that we are lazy. In fact, laziniss is a life wisdom for preserving strength. Is that smart?"

NO.13 · *The Giant Panda You Don't Know*

Fragrant Panda Feces

熊猫便便是香的

你知道大熊猫的便便为什么是香的吗?
边吃边排便,你注意过大熊猫的这一招吗?

Giant

Do you know why giant panda feces are fragrant? Have you ever noticed that the giant panda defecates while eating?

与众不同的熊猫便便

UNUSUAL GIANT PANDA FECES

便便什么味儿？这不是废话吗！当然是臭的！可你别不服气，人家大熊猫不仅颜值高、身体棒、人气旺，连拉的便便都与众不同——是香的！真是香的！

What do feces smell like? Needless to say, it's not pleasant! The giant panda, superior in appearance, health and popularity, defecates differently—fragrantly! It is true!

大熊猫具有特殊的消化系统。它们消化道短，采食竹子后8～9小时就能排出粪便，不像牛、马等植食动物，食物要在胃里储存、消化24小时以上才会排出。

成年大熊猫的粪便呈纺锤形，粪便中经常可以看到完整的竹叶、竹竿和竹枝。因为大部分竹子没有经过发酵就被排出来了，所以大熊猫的新鲜粪便一点儿也不臭，反而有一股淡淡的竹子的清香味儿。

Owing to their unique digestion system and short digestive tract, the giant pandas defecate 8-9 hours after ingesting bamboo. It is different from cattle, horses, and other herbivores that defecate after storing and digesting food in the stomach for over 24 hours.

From the fusiform feces of adult giant pandas, complete bamboo leaves, stems, and branches are often observed. As most bamboo is defecated without being fermented, the fresh feces is accompanied by a faint delicate fragrance of bamboo instead of stink.

摄影：魏辅文
Photo by Wei Fuwen

新鲜的熊猫便便富含植物纤维
Fresh panda feces is rich in plant fiber

JUDGING HEALTH BY FECES

As different giant pandas eat different lengths of bamboo fragment every time, the lengths of bamboo fragment fibers left in the feces are different. Sometimes they eat different bamboo, and the color of feces is also different.

If the giant panda eats bamboo shoots, the feces are pale yellow; if they eat bamboo stems, the feces are yellow-green; if they eat bamboo leaves, the feces are green. Researchers often judge their ages and health conditions by the color of feces and the length of residual bamboo fragment fibers and analyze their intestinal flora structure by researching their feces.

观便便，知健康

不同的大熊猫在进食时每次咬掉的竹节的长短不同，所以残留在粪便中的竹节纤维的长度也不同。又因为吃的竹子的部位不同，它们粪便的颜色也会不同。

如果吃的是竹笋，粪便呈淡黄色；如果吃的是竹竿，粪便呈黄绿色；如果吃的是竹叶，粪便呈绿色。研究人员常常根据粪便的颜色，其中残留的竹节纤维的长短来判断大熊猫的年龄和健康状况；还会通过研究粪便的成分，探析大熊猫肠道内的菌群组成。

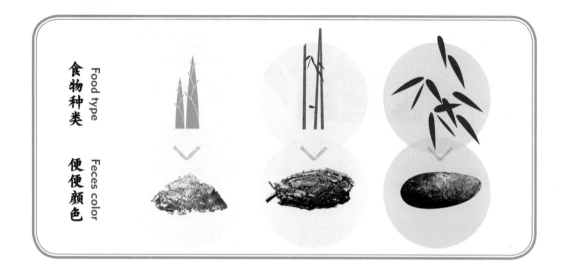

熊猫便便也有用

GIANT PANDA FECES ARE USEFUL

大熊猫的便便可以用来造纸，这是真的！

成年大熊猫吃竹笋时每日的排便量在20千克左右，有些能达到40千克；吃竹子时每日的排便量约10~20千克。

它们的粪便可以回收加工成纸张或工艺品出售，还可以作为肥料用来种植农作物。

Giant panda feces can be used to make paper. It's true!

The adult giant pandas defecate a daily quantity of about 20 kg (40 kg for some of them) and about 10-20 kg when they ingest bamboo shoot and bamboo, respectively.

The feces can be recycled and processed as paper or works of art for sale, as well as used as fertilizer for crops.

大熊猫小百科

熊猫便便纸，来一张？

大熊猫在消化竹子的时候只吸收了竹子中的糖分、淀粉和水分，竹子中的植物纤维并没有被消化掉。竹子本就是造纸的重要原料，人类在用竹子造纸的时候，也要将其中的糖分和水分去掉，利用其中的植物纤维来制造纸张。大熊猫的这一消化过程节省了造纸成本，也在一定程度上减轻了垃圾处理的压力。

GIANT PANDA TIDBITS

Would You Like a Piece of Paper Made From Giant Panda Feces?

During the digestion process, giant pandas digest only the sugar, starch and moisture instead of the plant fibers. Bamboo is a key material for papermaking. When making paper from bamboos, we remove the sugar and moisture and use the plant fibers. However, the digestive process of giant pandas saves manufacturing costs and adequately relieves the pressure of garbage disposal.

Neglected

NO.14 · The Giant Panda You Don't Know

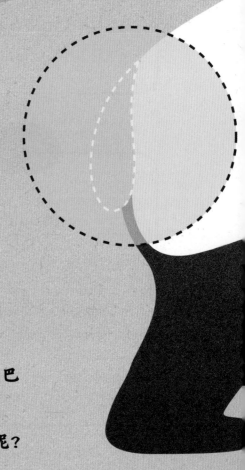

被忽略的尾巴

大熊猫也有尾巴。在你的印象里,它们的尾巴是什么颜色的呢?

在大熊猫的成长过程中,尾巴会有什么变化呢?

Tail

Giant pandas also have tails. In your impression, what is the color of the tails? How will the tail of the giant panda changes during its growth process?

大熊猫有尾巴吗？

这个问题可能会让好多人发愣：还真没注意过呀……其实，大熊猫是有尾巴的，而且，它们从小到大，尾巴的颜色和粗细的变化还不小呢。让我来考考你：大熊猫的尾巴是黑色的还是白色的呢？

DO GIANT PANDAS HAVE TAILS?

It may be an astonishing question since many of us might have not noticed. Actually, giant pandas have tails, which are quite different in terms of color and thickness from their births to adulthood. Can you guess that the tail of giant panda is black or white?

黑尾巴？ Black tail?

白尾巴？ White tail?

THE GIANT PANDA'S TAIL CHANGES IN DIFFERENT GROWTH STAGES

大熊猫的尾巴在不同成长阶段的变化

才出生时它们的尾巴比较细,是粉色的,表面还有稀疏的白毛。那时候它们小小的身体很能凸显尾巴的存在,尾巴的长度大概是身体长度的1/4。

When they are newborn, their tails are relatively thin and pink with sparse white fur dotted across the skin. At that time, the tails are a large part of their small bodies, accounting for 1/4 of the body length.

刚出生时
At birth

它们出生2个月左右,尾巴长出了浓密的白毛。

Two months later after birth, their tails have sprouted thick white fur.

出生2个月左右
About 2 months old

随着年龄的增长，相对于身体来讲，它们尾巴的长度增长非常有限，但是宽度有所增加。

As they grow older, their tails grow less long but thick relative to their bodies.

半岁左右
About half a year old

成年后
Adulthood

成年大熊猫的尾巴是白色的、扁平的，相对于庞大的身躯，它们的尾巴显得非常短小。这时它们尾巴的长度还不到全身长度的1/10，在胖嘟嘟的身体上，相当不起眼，以致于有时候让人忽略了尾巴的存在。

The white, flat tails of adult giant pandas are quite short compared to their big bodies. One tenth of their body length, the tails are so inconspicuous when compared with their chubby bodies that they sometimes go unnoticed.

大熊猫小百科

尾巴式信号发射器

虽然平时大熊猫的尾巴总是低垂着紧贴臀部,并不能像某些动物(猫、狗)一样用摆动尾巴来表达情绪,但是在它们的世界里,尾巴占据着相当重要的位置。它们的尾巴基部有一块无毛的裸露区,这就是肛周腺所在的地方。它们会用扁平如刷子的尾巴涂抹肛周腺分泌物来进行气味标记,从而实现与其他大熊猫的沟通。大熊猫小小的尾巴,在家族的文明体系建设中发挥着大大的作用。无声的沟通,堪比有声的交流。

GIANT PANDA TIDBITS

Tail as a Signal Emitter

Giant pandas usually keep their tails down, close to their buttocks, rather than swinging them like other animals (cats and dogs) to express their emotions. However, their tails play a very important role in their world. At the root of their tails is a furless bare area for circumanal glands. They use their brush-like flat tails to smear secretions from the circumanal glands as scent marks, actualizing communication with other giant pandas. The tiny giant panda tails have played a key role in constructing a communication system where silence is comparable to words.

The Giant Panda You Don't Know

NO.15

Giant "Signal Source"

大熊猫的"信号源"

雄性大熊猫为什么倒立着尿尿?
喜欢独居的野生大熊猫一般什么时候约会呢?

Panda's

Why do male giant pandas stand on its head to discharge urine? When do giant pandas that prefer to live solitary in the wild date?

大熊猫是独居动物

大熊猫其实一点儿也不喜欢集体生活。野外的大熊猫一般选择独居，生活在高山上薄雾弥漫的茂密竹林中。它们有自己的专属领地，彼此之间很少直接接触。

GIANT PANDA IS A SOLITARY ANIMAL

In fact, giant pandas do not prefer living in groups at all. Those in the wild usually live alone in bamboo forests surrounded by high mountains and spreading mists. They hardly ever meet each other directly due to having their own territories.

野生大熊猫出没的山林
Forest frequented by wild giant pandas

NO.15 大熊猫的"信号源"

气味标记是大熊猫主要的交流方式

因为长期生活在茂密的竹林中,大熊猫的视觉功能派不上大用场,相对较弱,所以它们的视力并不怎么好,但是嗅觉和听觉非常厉害。

气味标记是大熊猫最主要的交流方式。它们会在自然环境里留下分泌物和排泄物来标记资源和领地,其他大熊猫闻到这些气味后,知道这个区域已经被占领了,为了避免不必要的争斗,就会离开去寻找新的领地。

每当春暖花开,就到了它们谈恋爱的季节。一只雌性大熊猫的气味标记可能表示它已经做好谈恋爱的

闻气味 Smelling

SCENT MARKS ARE THEIR PRIMARY METHOD OF COMMUNICATION

As giant pandas live in dense bamboo forests for a long time and hardly exert their visual functions, they have relatively inferior vision but excellent senses of smell and hearing.

Scent marks are the primary method of communication. They leave secretion and feces to mark resources and territories. When other giant pandas smell those scents, they know that the territory is occupied and usually leave for another territory to avoid unnecessary fights.

When the spring comes, they will fall in love. If a female giant panda marks its scent, it probably means that she is

做标记
Leaving scent marks on tree trunks

准备，雄性大熊猫可以通过这种特殊的"信号源"找到它。

雄性大熊猫倒立着尿尿，为的是能够把尿液排到更高的地方。位置越高，意味着雄性大熊猫的体形越大，就越能获得雌性大熊猫的青睐。不是谈恋爱的季节，一闻到陌生大熊猫的气味，它们就会彼此走开，毕竟它们都有自己的领地。

所以，这些气味标记能让大熊猫们互相回避或是聚集到一起，够神奇吧！

prepared to fall in love, and male giant pandas can find her by her special "signal source".

Male giant pandas stand on their heads to urinate from a higher position. Urinating from a higher position indicates a larger giant panda, which is more attractive to the female giant panda. Except for the spring, they will leave after smelling unknown giant pandas as they have their own living area after all.

Therefore, the scent marks are a magical way for giant pandas to stay away from each other or gather.

做标记
Leaving scent marks on tree trunks

摄影：雍严格
Photo by Yong Yan'ge

大熊猫小百科

气味标记方式多

大熊猫经常用尿液、肛周腺分泌物或二者的混合物来做标记，将气味涂在柱子、树桩、墙上、地上，以及它们经常经过的地方。它们做标记的时候，会晃动头部，嘴巴半张。做了标记以后，它们会用剥掉树皮、留下抓痕等方式，来引起其他大熊猫的注意。

GIANT PANDA TIDBITS

Varieties of Scent Marks

Giant pandas usually smear their urine, circumanal gland secretions or a mixture of both on the pillars, tree trunks, walls, the ground, and at the places they often visit. When leaving their marks, they shake their heads with their mouths half-open. Afterwards, they peel off barks or leave scratches at marked places to garner attention from other giant pandas.

NO.16 The Giant Panda You Don't Know

Giant Growth

16

"滚滚"成长记

大熊猫宝宝出生后多少天才会睁开眼睛呢?

出生时100多克,成年后100多千克,你见过体重变化如此大的动物吗?

Panda's Process

How many days will giant panda cubs finally open their eyes after birth? Have you ever seen an animal that changes weight as significantly as it grows up, from over 100 g at birth to over 100 kg?

可爱的大熊猫宝宝们

THE LOVELY BABY PANDAS

大家对家养宠物猫、狗的成长变化非常熟悉，见过它们才出生时的样子，也经常能在马路边、公园里、小区里见到它们的身影。但是，我们的国宝大熊猫可不是随便在哪里都能看到的，大熊猫宝宝也不是什么时候都可以看到的。

We are familiar with how pet cats and dogs grow. We know how they look like at birth, and often see them by roadside, in parks, and in neighborhoods. However, it is unlikely we will see the national treasure that is the giant pandas anywhere and their cubs even less.

新生宝宝

New-born Cub

From June to September of each year, giant panda cubs are born.

The newborn giant panda cubs with pink skin and sparse white fur weigh around 120 g on average, which is about 1/1000 of their mothers. Assuming that an infant weighs 3 kg at birth while its mother weighs about 3000 kg. It's amazing!

每年的6~9月,是大熊猫宝宝出生的季节。

新生大熊猫宝宝的皮肤是粉色的,上面长着稀疏的白毛。它们的平均体重约为120克,只有妈妈体重的1/1000左右。我们人类小宝宝出生时的体重一般在3千克左右,按照这个比值计算,那妈妈的体重应该达到3000千克!太惊人了!

大熊猫的孕期很特别——忽长忽短不固定。在野外，大熊猫的孕期平均为144天，最短68天，最长200天。圈养条件下，大熊猫怀单胎的孕期平均为142天，怀双胞胎的孕期平均为132天，在日本甚至出现过怀孕324天的特例。

刚出生时，大熊猫宝宝的各个器官都没有发育完全，免疫力很低。这时候，妈妈的母乳给大熊猫宝宝提供了必需的营养和免疫抗体。有妈妈可真好！

Pregnancies for giant pandas are different-changeable and unfixed. Giant pandas in the wild have a pregnancy period of about 144 days on average (68-200 days), while captive giant pandas have a pregnancy period of 142 days for one cub and 132 days for twins on average. However, a 324 days' pregnancy occur in Japan.

At birth, their organs are not fully developed, and their immune system is weak. At this time, the breast milk from panda mothers provides necessary nutrition and immune antibody.

人工喂养初生大熊猫宝宝
Artificially feeding newborn panda cub

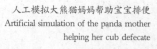

大熊猫初乳
Giant panda colostrum

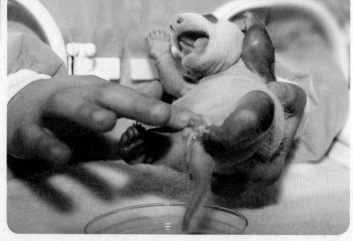

人工模拟大熊猫妈妈帮助宝宝排便
Artificial simulation of the panda mother helping her cub defecate

2 周

出生2周左右，黑色的毛开始生长，眼睛、耳朵、眼眶、四肢和肩带渐渐变黑，它们开始变得黑白分明。

Two Weeks
About 2 weeks after birth, their black fur begins to grow. When their eyes, ears, eye sockets, limbs, and shoulders gradually darken, they are fully black and white.

30 天

30天左右，它们长出了黑白相间的毛，变得更像妈妈的模样了。

One Month
In about 30 days, they look more like their mothers as the black and white fur grows in.

40 多天

40多天时，大熊猫宝宝终于睁开眼睛了，但这个时候它们还看不见，因为它们的眼睛里有一层虹膜。2个多月的时候，它们就能真正看清这个新奇的世界啦。而在这个时候，它们也开始慢慢地爬动，随后开始学习走路。

Over Forty Days
After 40 days, they open their eyes at last, but they can't see because of an iris layer in their eyes. At about 2 months, they can truly see the novel world. At this time, they begin to crawl slowly and then learn to walk.

半岁到1岁

半岁到1岁是大熊猫宝宝最活泼的时候,它们开始在活动场玩耍和学习各种生存技能:奔跑,翻滚,嬉闹,爬树,学吃竹子……

A Half to One Year

At about 6-12 months, they are the liveliest, beginning to play on the playground while learning all necessary life skills: running, rolling, frolicking, climbing trees, eating bamboo…

1岁半到2岁

从1岁半到2岁,它们长齐了所有牙齿。断奶后,它们就会离开妈妈,开始独立生活。

One Year and a Half to Two Years
From 1.5 to 2 years old, they have all their teeth and leave their mothers after weaning to live independently.

亚成年阶段

大熊猫断奶后到成年前的亚成年阶段，相当于人类的青少年时期。这一阶段，大熊猫的食物从以营养丰富的奶为主逐渐过渡到以高纤维、低营养的竹子为主。它们在生理和心理上也逐渐成熟了。

Sub-adult Period

From weaning period to their sub-adult stage, which is equivalent to the adolescent period of human, giant pandas gradually mature physiologically and physically with their staple food transiting from nutritious milk to high-fiber and low-nutrition bamboo.

成年期

在5岁左右，大熊猫进入成年期，可以开始繁育后代了。这个时期，它们的身长能达到160～180厘米，体重能达到80～150千克。

在野外，大熊猫的寿命一般为18～20岁；在圈养条件下，由于有饲养员的精心照料，大熊猫的寿命一般为25岁左右，有的甚至能达到30多岁。

这就是大熊猫的成长过程，你了解了吗？

Adult Period

From around 5 years old, they enter into adulthood and begin to breed cubs. During this period, they may reach a height of 160-180 cm and a weight of 80-150 kg.

In the wild, giant pandas live 18-20 years on average; under the conditions of captive breeding, giant pandas live 25 years in general and some of them even live to their thirties thanks to the excellent care from keepers.

That is how giant pandas grow. Do you understand?

FUN FACTS
趣闻一串串

创下两项吉尼斯纪录的最长寿大熊猫

大熊猫"佳佳"1978年出生在四川省青川县。为庆祝香港回归2周年,"佳佳"于1999年被赠送给香港特别行政区,生活在香港海洋公园。2015年7月28日,香港海洋公园为"佳佳"庆祝了37岁生日。当天,吉尼斯世界纪录认证官宣布,"佳佳"刷新了两项吉尼斯世界纪录,成为"迄今为止最长寿的圈养大熊猫"及"最长寿的在世圈养大熊猫"。大熊猫37岁的年纪相当于人类110岁高龄。后来,"佳佳"的健康状况急剧恶化,清醒的时间越来越少,对食物没有兴趣,甚至无法走动。2016年10月16日,香港特别行政区政府渔护署及香港海洋公园的兽医在为"佳佳"诊治后,一致同意为它进行安乐死。2016年,最长寿的大熊猫去世,享年38岁。

The Most Long-living Giant Panda With Two Guinness Records

Born in 1978 in Qingchuan, Sichuan, the giant panda Jia Jia was presented to the Hong Kong Special Administrative Region in 1999 to celebrate the second anniversary of Hong Kong's Return to China, and lived in the Ocean Park Hong Kong. On July 28, 2015, the Park celebrated Jia Jia's 37th birthday. On that day, the authenticator from Guinness World Records declared that Jia

Jia set two Guinness records, one for the Oldest Giant Panda in Captivity and the other for the Oldest Living Giant Panda in Captivity. A 37-year-old giant panda is equivalent to a 110-year-old human. Later on, Jia Jia's conditions worsened, staying awake less and less, losing interest in food and even failing to move. On October 16, 2016, veterinarians from the Hong Kong Special Administrative Region Agriculture, Fisheries and Conservation Department and the Ocean Park Hong Kong unanimously agreed to euthanize Jia Jia after diagnosis and treatment. In 2016, the oldest living giant panda passed away at the age of 38.

还有比我更长寿的大熊猫吗?
Are there any giant pandas that have lived longer than me?

FUN FACTS
趣闻一串串

初生体重最轻的大熊猫宝宝

2019年6月11日，大熊猫"成大"产下双胞胎姐妹花"成风"和"成浪"，其中"成浪"的初生体重仅有42.8克，脱水后只有37.9克，不到一个鸡蛋的重量。"成浪"出生时，在场的专家和饲养员又紧张又惊讶，因为它的体重只有正常初生大熊猫宝宝的四分之一。由于个体太小，"成浪"无法在妈妈那里吃到初乳，科研人员就想办法从它妈妈那里挤来初乳，为它人工补奶。到1岁时，"成浪"的体重已经长到了38千克（姐姐"成风"体重为39千克），达到同龄大熊猫正常体重。

The Giant Panda Cub With the Lightest Birth Weight

On June 11, 2019, the giant panda Cheng Da gave birth to the twin girls Cheng Feng and Cheng Lang. Cheng Lang, with a birth weight of 42.8 g (37.9 g after dehydration, less than the mass of an egg) . At Cheng Lang's birth, experts and keepers were nervous and surprised that she weighed only a quarter of what is a normal weight for a newborn giant panda. As Cheng Lang was too small to drink breast milk from her mother, researchers worked to get breast milk from her mother and feed her manually. At the age of 1, Cheng Lang weighed 38 kg (the weight of her elder sister was 39 kg), weighing the same as a normal giant panda.

NO.16 Giant Panda's Growth Process NO.16 "滚滚"成长记

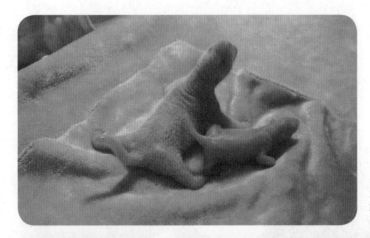

刚出生的姐妹花"成凤"和"成浪"
Newborn female giant panda twins Cheng Feng and Cheng Lang

"我们长大啦!"
"We become bigger!"

The Giant Panda You Don't Know

NO.17

The Story Giant Xiao Man

大熊猫小满的故事

你知道大熊猫宝宝在小的时候不会自己排便吗？
在野外，大熊猫妈妈生了双胞胎宝宝后会做出怎样的选择呢？

# of Panda

Do you know giant panda cubs are not able to defecate on their own?
What decision the giant panda mother will make when that delivering twins in the wild?

嗨，我是小满！

在野外，每年秋天是大熊猫妈妈的生产季节。大熊猫妈妈会选择一个隐蔽的树洞或一个天然的岩洞，在里面铺垫树枝和干草作为巢穴，用来哺育即将出生的宝宝。

这是大熊猫小满，这是它与妈妈的故事。

摄影：雍严格
Photo by Yong Yan'ge

HI, I AM XIAO MAN!

Each autumn (delivery season), giant panda mothers in the wild will choose a hidden tree den or a natural-forming cave with branches and dried bushes laid out to look after the upcoming cubs.

This story is about the giant panda Xiao Man and his mother.

"It was told by my mother that she held me in her arms to warm and protect me everywhere all the time for a few weeks after I was born. When she moved, she held me in her mouth and kept me where she was.

"Since my eyes were closed at birth, I uttered different cries to get my mother's attention when I was hungry, needed to defecate, or suffered from cold, heat, or other discomfort. At this age, cries were an extremely important tool to communicate with my mother.

"As newborn giant panda cubs are incapable of defecating or urinating independently, their mothers lick their perineum with their tongue to help them by stimulating their bowel movements. That's how my mother helps me poop.

"妈妈说,我刚出生那几周,她一直把我抱在怀里,温暖我、保护我,几乎寸步不离。她需要移动的时候就会把我衔在嘴里,总之她在哪儿我就在哪儿。

"我才出生时还没有睁开眼睛,如果饿了、想排便了、冷了热了,或者有任何的不舒服,我就会发出不同的声音来提醒妈妈。这时候,我的叫声是我和妈妈之间非常重要的交流工具。

"由于才出生的大熊猫宝宝都还没有自主排出大、小便的能力,大熊猫妈妈会用舌头舔宝宝的会阴部,通过这样的刺激来帮助宝宝排便。我的妈妈也是这样帮助我排便的。

摄影:张志和
Photo by Zhang Zhihe

"爬树好难啊,可是妈妈超级有耐心。"
"It's so hard to climb, but my mother is super patient."

"妈妈当年只生了我一个孩子。但是,妈妈告诉我,如果在野外,大熊猫妈妈生的是双胞胎的话,她们一般会放弃身体弱小的那一只,而选择身体强壮的那只来抚养,因为这样能保证在野外艰苦的环境下至少有一个宝宝可以存活。我很理解妈妈们的选择。

"我现在已经1岁半了,妈妈在这段时间里一直在照顾我,所以这1年半里妈妈都没有再生宝宝。妈妈最多会照顾我到2岁,之后我就要独自去生活了,而妈妈也要再次准备为大熊猫家族繁衍后代。

"I was the only litter when my mother gave birth to me. However, my mother told me that if giant panda mothers give birth to twins in the wild, they choose the stronger one to raise and discard the weaker one to ensure at least one cub could survive a hard and wild environment. I understand the choice of the giant panda mothers.

"Now I am 1.5 years old. During this period, my mother hasn't had any other babies but has looked after me all. As soon as I am 2 years old at most, I will live independently and my mother will get ready to have babies again.

"Living with our mothers is such an important period for us giant pandas. My

"我能行!"
"I can do it!"

mother has taught me many life skills, such as how to find food and water in the wild, escape from predators and fight diseases, court and give birth, and be a real giant panda. It is almost the only chance for us giant pandas to live with another of our kind and to learn.

"I'll cherish the days with my mother and learn all skills to get ready to live by myself."

"和妈妈在一起的这段时间，对我们大熊猫来说是非常重要的。妈妈教会了我很多生存技能，比如怎样在野外寻找食物和水源，怎样躲避天敌，怎样抵御疾病，怎样谈情说爱、生儿育女，怎样成为一只真正的大熊猫。这几乎是大熊猫一生中唯一一次和另外一只大熊猫长时间共处，也几乎是唯一一次系统学习的机会。

"我会好好珍惜和妈妈在一起的日子，学会各种本领，为独自闯荡世界做好准备。"

"我爱你，妈妈！"
"I love you, mom!"

大熊猫小百科

巧妙的换仔技术

在圈养条件下，当大熊猫妈妈生了双胞胎时，为了将宝宝都抚育成活，饲养员会取走一只，让大熊猫妈妈仅带一个宝宝。取走的宝宝将被放在育婴箱中，由饲养员们照顾。育婴箱中的温度与湿度和大熊猫妈妈怀中的温度与湿度一样。当育婴箱中的宝宝饿了，饲养员将会取走大熊猫妈妈怀中的宝宝，将育婴箱中的宝宝交还给大熊猫妈妈让其喂奶，同时让宝宝享受妈妈的爱，就这样不停地交换，让两只大熊猫宝宝都能成活。

GIANT PANDA TIDBITS

Clever Cub Swapping Technique

When the giant panda mother has twins, to ensure that both of them survive, the keeper will take one away into the incubator, where the temperature and moisture are the same as those of the mother's arms, and looked after by keepers, while the other is left under the care of the mother. When the giant panda cub in the incubator is hungry, the keeper will take away the cub in the mother's arms and give the one in the incubator to the panda mother to feed on breast milk and enjoy their mother's love. This allows both giant panda cubs to survive.

换仔技术的运用，帮助大熊猫妈妈成功养育了双胞胎宝宝
The cub swapping technique has helped giant panda mothers successfully raise twins

NO.18 The Giant Panda You Don't Know

Giant Language

教你听懂"熊猫语"

为什么大熊猫的叫声有时候像狗吠，有时候又像鸟鸣？
你知道13种已经解码的"熊猫语"分别代表什么意思吗？

Panda

Why do giant panda barks and chirps? Do you know what the 13 decoded giant panda vocalizations mean?

大熊猫也有自己的语言

人类之间靠语言进行交流，动物之间也有自己的交流方式。

平时"沉默寡言"的大熊猫也不例外，它们会在不同时期、不同环境、不同情况下用独特的"熊猫语"进行交流。

研究人员通过对大熊猫的叫声进行收集，利用声谱特征分析，又参照它们的行为表现，从而分析出它们的叫声所表达的意义。研究发现大熊猫的叫声多达10多种，有的用于母子之间交流，有的用于求偶，有

"我饿了！" "I'm hungry!"

"你好！" "Hello!"

GIANT PANDAS ALSO HAVE THEIR OWN LANGUAGE

Humans communicate through language, and animals have their own ways of communicating.

The same holds true for the often-silent giant pandas, who exchange information with their unique "language" based on different periods, environments, and situations.

Researchers decoded giant panda cries by collecting them, analyzing their spectrum characteristics, and referencing giant panda behaviors. Studies concluded that giant pandas can utter more than ten kinds of cries under different scenarios, including mother-child communication,

"我可爱吗？" "Am I adorable?"

courtship, and deterrence or warning.

Giant panda cubs can make relatively sharp vocalizations like squeaks, bowwow, and hoots at birth to attract their mothers' attention when they are hungry, uncomfortable, or frightened and want to be comforted by their mother.

The pandas can utter more vocalizations as they get older. They scream when they are frightened, roar when angry, and hum when enjoying something.

的用于表达威慑或警告。

　　大熊猫宝宝才出生就能发出吱吱、哇哇、咕咕的叫声，而且声音比较尖锐。当它们饿了、身体不舒服或者受到了惊吓，想得到妈妈的安慰的时候，会发出叫声，以吸引妈妈的注意力。

　　随着年龄增长，大熊猫能够发出的声音也越来越多。当它们受到惊吓的时候，会发出尖叫声。生气时，它们会发出怒吼声。而当它们感到舒服的时候，则会发出酥软的哼哼声。

到了谈恋爱的季节,雄性大熊猫会发出犬吠声。这是在向周围的雄性大熊猫发出警告:"离我远点儿,别惹我!"也是在向雌性大熊猫传达爱意:"看我有多强大!"

当它们双双坠入爱河,雄性大熊猫会发出咩叫声,雌性大熊猫的叫声则会由咩叫声变为类似于鸟鸣的唧唧声。

已被解码的大熊猫的13种叫声,能帮助研究人员进一步了解它们的生活习性和心理,走进它们的世界。不过,要想真正全面地了解大熊猫,我们还需要继续努力探索。

During estrus, the male giant pandas would bark, warning other males in the vicinity to "stay away and don't mess with me" and attracting females to "see how strong I am".

When two pandas fall in love, the male bleats while the female changes from bleats to a sound that resembles chirps.

The 13 types of vocalizations that have been decoded can assist researchers to further understand giant panda habits and psychology and enter their world. However, more endeavors are needed to fully understand them.

"离我远点儿!" "Stay away from me!"

大熊猫小百科

大熊猫不同叫声的含义

嗯叫：幼仔寻母　　　　歇叫：拒绝、厌烦

哇哇叫：幼仔不适、饥饿　　犬叫和嗷叫：打斗、警告

尖叫：害怕、惊恐　　　　咩叫：发情、求偶

GIANT PANDA TIDBITS

Meanings of Different Giant Panda Cries

Hum：call for mother

Snort：reject, annoy

Bowwow：uncomfortable, hungry

Bark and Growl：fight, warn

Squeal：scared, frightened

Bleat：estrous, courtship

The Giant Panda You Don't Know

NO.19

Tough

难缠的疾病

强壮的大熊猫也难以摆脱病痛的困扰。
对它们的生命安全最具威胁的疾病主要有哪些呢？
你知道兽医是怎么给大熊猫喂药的吗？

Diseases

Even strong giant pandas can hardly escape illness. So what are the diseases that primarily threaten their lives? Do you know how veterinarians help giant pandas take medicine?

大熊猫也会生病

2020年初，一场突如其来的新冠疫情让我们的世界戛然"停摆"。历史上，人类一直在同疾病抗争。实际上，大熊猫也会像人类和其他动物一样，患上各种疾病。

GIANT PANDAS GET SICK, TOO

At the beginning of 2020, the COVID-19 pandemic swiftly swept the world and suspended how our society normally functions. The history of humankind is also the history of fighting diseases. In fact, just like humans or other animals, giant pandas suffer from a variety of diseases.

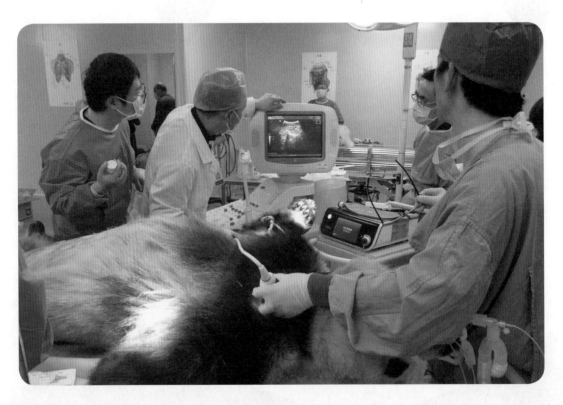

NO.19 Tough Diseases　　　　　NO.19 难缠的疾病

大熊猫的疾病通常分为内科疾病、外科疾病、产科疾病、传染性疾病和寄生虫病几大类。某些疾病一旦发生，可能会对它们的生命造成威胁，也可能会给整个大熊猫家族带来灭顶之灾。

Giant panda diseases are usually divided into medical diseases, surgical diseases, obstetric diseases, infectious diseases, and parasitic diseases. Certain diseases may threaten their lives and even extirpate to the entire species once they occur.

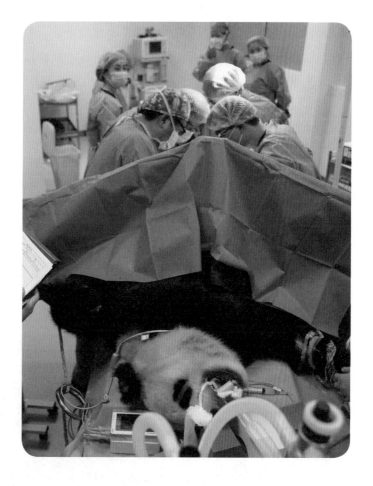

内科疾病

内科疾病范围广泛，包括消化系统疾病、呼吸系统疾病、泌尿系统疾病、神经系统疾病、营养代谢性疾病、中毒性疾病、造血系统疾病等。大熊猫最常见的内科疾病有急性胃肠炎、慢性胃肠炎、感冒、肺炎等。对它们的生命安全最具威胁的内科疾病主要包括中暑、肠梗阻、肠扭转、肠套叠、急性胰腺炎、肾功能衰竭等。

Medical Diseases

Medical diseases cover a wide range, involving digestive system diseases, respiratory system diseases, urinary system diseases, nervous system diseases, nutritional and metabolic diseases, toxic diseases, and hematopoietic system diseases. The most common medical diseases for giant pandas include acute gastroenteritis, chronic gastroenteritis, colds, and pneumonia, while the most life-threatening diseases include heatstroke, intestinal obstruction, volvulus, intussusception, acute pancreatitis, and kidney failure.

大熊猫"囡囡"
The giant panda Nan Nan

FUN FACTS

趣闻一串串

"胖美女"开刀取出了啥?

大熊猫"囡囡"胃口十分好。身材丰满、心宽体胖的它荣登"斧头山第一胖美女"宝座。但是,可爱的它也有肠梗阻的"黑历史"。2016年12月,"囡囡"出现烦躁不安、腹疼、频频举尾、排不出粪便等症状。经专家会诊,初步诊断为肠梗阻。通过手术,医生从"囡囡"的肠道中取出了堵塞的8个粪团,共重2.8千克。最终,"囡囡"在医护人员的精心护理下康复。"囡囡"开创了大熊猫"开刀取屎"的先例。

What did One Operation Remove From the "Plump Beauty"?

The giant panda Nan Nan has a very good appetite and as well as a plump body, which crowned her the Top Plump Beauty in Mount Futou. However, this charming beauty once suffered from an intestinal obstruction. In December 2016, Nan Nan manifested symptoms, including agitation, abdominal pain, and frequent tail lifting with a failure to defecate. After expert consultation and examination, the preliminary diagnosis was an intestinal obstruction. The operation removed 8 clogged fecal pellets totaling 2.8 kg from her intestinal tract, and she eventually recovered thanks to meticulous care from medical staff. In addition, Nan Nan also set a precedent for needing an operation to remove feces.

外科疾病

大熊猫最常见的外科疾病有外伤、骨折、颅脑损伤、肿瘤等，这些伤病都必须通过外科手术才能完成治疗。但是，由于大熊猫们没那么安分，会不停地抓扯伤口，从而影响伤口的愈合，所以它们外科疾病治疗成功的关键在于手术后的护理。

Surgical Diseases

The most common surgical diseases are trauma, fractures, head injuries, tumors, etc. These diseases must be treated with surgery. Since giant pandas may scratch the wound to hinder its healing, the key to treating surgical diseases is post-operative care.

大熊猫"毛豆"
The giant panda Mao Dou

怪不好意思的……
It's embarrassing...

FUN FACTS

趣闻一串串

"社会你豆哥" 阴沟里翻船

大熊猫"毛豆"性格霸道,爱跟小伙伴"斗殴",且战斗力很强,被大家称为"社会你豆哥"。不过再厉害的大熊猫,也有阴沟里翻船的时候。2018年3月,"毛豆"在和小伙伴打闹后,右后肢无法着地。通过一系列检查,被确诊为股骨骨折,医生当即为它进行了手术。在医护团队的精心护理下,"毛豆"的精神和食欲迅速恢复,伤口也恢复良好。最后,"毛豆"又回到了幼儿园的同级小伙伴中间。同时,它因为手术中后腿被剃掉毛发的"脱裤"照爆红网络。

"Gangster" Mao Dou Broke His Bones

The giant panda Mao Dou is a bully, who loves to pick fights with other giant pandas and shows a strong fighting ability, winning the title "gangster". However, even the strongest pandas get hurt. In March 2018, it was discovered that Mao Dou's right hind limb did not touch the ground after a fight with one of his friends. He was diagnosed with a femoral fracture after a series of examinations and was immediately prepped for surgery. Thanks to the meticulous care of the medical team, Mao Dou recovered well in spirit and appetite with his wounds fully healing. In the end, Mao Dou returned to the kindergarten and re-joined the class with his peers. Plus, he became an Internet sensation after his hind legs were shaved for the operation.

产科疾病

大熊猫的产科疾病主要包括假孕、流产、难产、外阴水肿、子宫内膜炎、卵巢囊肿、输卵管堵塞、不育症等，其中假孕、流产较为常见。大熊猫初生幼仔的体重只有大熊猫妈妈体重的1/1000，因为胎儿个体太小，发生难产的可能性极小；可能出现的情况是，大熊猫妈妈分娩时因子宫收缩无力而出现产力性难产。

Obstetric Diseases

Giant panda obstetric diseases primarily include pseudo-pregnancy, miscarriage, dystocia, vulvar edema, endometritis, ovarian cysts, blocked fallopian tubes, and infertility, among which, pseudo-pregnancy and abortion are common. A newborn cub weighs only 0.1% of its mother's weight, so it is extremely unlikely that panda mothers have dystocia because the fetus is too small. Dystocia usually occurs when the giant panda mother has weak uterine contractions during delivery.

大熊猫"二丫头"
The giant panda "Er Yatou"

"二丫头"在13岁时产下了自己的第一个宝宝，成功当上了母亲
Er Yatou, at the age of 13, gave birth to her first child and became a mother

FUN FACTS

趣闻一串串

"二丫头"险遇难产

2004年，大熊猫"二丫头"就要生宝宝了。但是，它在第一次破羊水后14小时内都没表现出任何产前反应。专家们通过B超检查发现，胎儿还一直停留在"二丫头"的子宫内。最后，专家们通过保守治疗方法，将已死亡的胎儿取出母体，避免了外科手术，最大程度地保护了"二丫头"。这是全球范围内报道的唯一一例大熊猫难产病例。

Giant Panda "Er Yatou" Suffered From Dystocia

In 2004, the giant panda "Er Yatou" was about to give birth, but she did not exhibit any prenatal reactions 14 hour after her first water broke. Experts conducted a B-ultrasound examination and found that the fetus remained in her uterus. In the end, experts adopted conservative treatment and discharged the dead fetus from the mother, avoiding a surgical operation and protecting the giant panda "Er Yatou" to the utmost extent. This is the only reported case of giant panda dystocia worldwide.

寄生虫病

大熊猫身上的寄生虫包括体内寄生虫和体外寄生虫两大类。目前，科学家们已在大熊猫身上发现了22种寄生虫，最为常见的寄生虫病包括蛔虫病、痒螨病、蠕形螨病、蜱病等。

Parasitic Diseases

Giant panda parasites include endoparasites and ectoparasites. So far, scientists have found 22 kinds of parasites, and the most common parasitic diseases primarily include ascariasis, psoroptic mange, demodicosis, and ixodism.

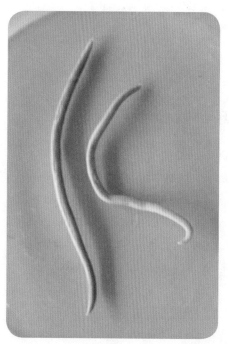

蛔虫
Roundworms

熊猫痒螨
Giant panda psoroptidae

锐跗硬蜱
Ixodes acutitarsus

FUN FACTS

趣闻一串串

它感染了1605条蛔虫！

在圈养条件下，大熊猫受到人类良好的医疗护理，寄生虫病的感染率和感染强度往往很低，通常不会对生命安全构成严重威胁。而在野生条件下，大熊猫无法得到来自人类的医疗护理，寄生虫病的感染率和感染强度都远高于圈养大熊猫，这已成为野生大熊猫消瘦和死亡的主要原因。成都大熊猫繁育研究基地一直致力于野外大熊猫的抢救工作，其抢救过的野外大熊猫蛔虫感染率几乎达到100%，曾在一只得到救治的野生大熊猫体内发现了1605条蛔虫。

Giant Panda Infected With 1,605 Roundworms!

In captivity, infection rates and intensities of parasitic diseases are low and do not pose a serious threat to giant pandas due to the excellent medical care provided by humans. However, giant pandas in the wild cannot receive any medical care, so the infection rates and intensities are much higher than that of captive giant pandas and have become the main causes of weight loss and death among wild giant pandas. The Chengdu Research Base of Giant Panda Breeding is committed to rescuing wild giant pandas and discovered that the giant panda roundworm infection rate in the wild fundamentally reaches 100% every time. A rescued giant panda in the wild hosted up to 1,605 roundworms.

传染性疾病

大熊猫的传染性疾病分为细菌性疾病、病毒性疾病、真菌性疾病。这些都是对它们的种群健康和生命安全具有极大威胁的疾病。每当一种未完全被科学家认识的新型传染病来袭，往往会给整个大熊猫圈养种群造成非常严重的损失，甚至带来毁灭性的打击。

Infectious Diseases

Giant panda infectious diseases are divided into bacterial diseases, viral diseases, and fungal diseases and pose a crucial threat to the health and safety of giant panda populations. The revelation of a new type of infectious disease before being fully recognized by scientists is often impossible to prevent and may cause severe losses or even crushing blows to the entire captive giant panda population.

FUN FACTS

趣闻一串串

可怕的大肠杆菌 O152

20世纪80年代因为抢救野生大熊猫而传入圈养种群的由侵袭性大肠杆菌O152引起的出血性肠炎，2年内导致了近20只大熊猫死亡。轮状病毒导致的大熊猫断奶幼兽顽固性腹泻病和慢性营养不良综合征，犬瘟热病毒导致的大熊猫瘟热病，细小病毒导致的大熊猫病毒性肠炎等，都曾给整个圈养种群造成过极大伤害。

经过全世界科学家的共同努力，我们已经成功地找到了以上病症的有效治疗方法和预防措施，并已研制出专门用于预防大熊猫犬瘟热病毒感染的疫苗。

Horrible E. coli O152

Hemorrhagic enteritis caused by the invasive E. coli O152, which was introduced into captive populations in the 1980s from rescued wild pandas, was responsible for nearly 20 deaths over 2 years. Intractable diarrhea and chronic malnutrition syndrome caused by rotavirus in weaning cubs, panda distemper caused by canine distemper virus, and panda viral enteritis caused by parvovirus have greatly harmed the entire captive population.

Thanks to the coordinated efforts of scientists around the world, we have successfully found effective treatment and prevention measures for these diseases, and screened out preventive vaccines specifically for pandas to fight canine distemper virus infection.

FUN FACTS

趣闻一串串

给滚滚喂药是个技术活

你都不爱吃药,大熊猫会乖乖吃药吗?当然不会啦!成年大熊猫个子高、体重大,关键还力大无穷,逼它们吃药简直不可能,但是兽医们可会想办法啦:他们会先根据大熊猫的年龄、体重、病情等调节用药剂量,然后根据药的形态和口感,加入适量的糖,再偷偷把药混合在牛奶和水中,或夹在窝窝头、水果中。"嗯嗯嗯,今天的牛奶有点儿甜,好喝!"你看,"滚滚"们咕咚咕咚、开开心心地喝了个精光。

Techniques to Get Giant Pandas to Take Their Medicine

Since you do not like taking medicine, will pandas voluntarily take medicine? Of course...not! The adult giant panda is tall, heavy, and strong, making it extremely difficult to force them to take their medicine. However, smart veterinarians have their own techniques: they will first adjust the dosage according to the giant panda's age, weight, and condition, then add appropriate sugar according to the shape and taste of the medicine, and secretly mix it into milk, water, bread, or fruits. "Hmm, today's milk is a bit sweet! Yummy!" You see, they finished it all gladly.

NO.19 Tough Diseases NO.19 难缠的疾病

兽医将药混在盆盆奶中
The veterinarian mixed medicine into the panda milk

"今天的窝窝头味道有些不一样……"
"This panda cake tastes a bit different today..."

NO.20

The Giant Panda You Don't Know

Be a Giant Guardian

来做大熊猫小卫士

你知道拒绝购买野生动物制品也是保护野生动物的一种方式吗？
节约用纸也能为保护大熊猫出一份力？

Panda

Do you know that refusing wildlife products is also a way to protect wild animals? Does saving paper contribute to giant panda conservation?

保护大熊猫
就是在保护我们自己

经常有人问，花那么多钱那么多精力保护大熊猫和其他野生动物，值得吗？华南植物园一位昆虫学博士说得好：这些比人类古老得多的野生动物在自然进化中早已和周边的生态环境形成了千丝万缕的共生关系。

"这就像一张蜘蛛网。一个物种灭绝了，就会出现一个空洞的点，而以这个点为起点，又会再次引发

PROTECT GIANT PANDAS IN FACT PROTECTS OURSELVES

Is it worthwhile to invest so much money and energy to protect giant pandas and other wild animals? To address this often-asked question, I would like to cite a doctor of entomology from the South China Botanical Garden, Chinese Academy of Science on wild animals, which are much older than humans are, forming an inseparable symbiotic relationship with surrounding ecological environment during natural evolution.

大熊猫的家园也是其他动物的家园
The giant panda habitat is also home to other specie

"It's like a spider web. When one species dies out, a hollow spot appears on the web, causing other spots to disappear... What you see is the extinction of one species, but what you don't see may be the collapse of an ecosystem."

For example, if the wild South China tiger becomes extinct, there must be a sharp decline in the deer and wild boar populations on which it feeds, triggering a series of ecological reactions. Since we humans are also a part of nature, are not maintaining the original ecological balance and protecting animals and plants on earth in fact protecting ourselves?

其他点的消失……你眼睛看得到的，是一个物种的灭绝；你看不到的，可能就是一片生态系统的崩塌。"

比如，华南虎野外种群消失了，其赖以为生的鹿科动物、野猪等也急剧减少，并由此引发了一连串的生态反应。我们人类也是大自然的一部分，尽可能地维持地球原生生态平衡、保护动植物，其实不就是在保护我们自己吗？

摄影：雍严格

Photo by Yong Yan'ge

野生动物的代言人

SPOKESPERSON FOR WILDLIFE

百年来，大熊猫作为"旗舰物种"和"伞护种"，在有效拯救和保护珍稀濒危物种、促进生物多样性的行动中发挥了重要作用。

今天，为实现对自然资源的科学保护和合理利用，国家批准设立了大熊猫国家公园。大熊猫国家公园的试点区规划范围包括四川省岷山片区、邛崃山-大相岭片区、陕西省秦岭片区和甘肃省白水江片区，整合了各类自然保护地80余个，区域内生活着1631只野生大熊猫（占全国野生种群数量的87.5%），面积27134平方千米（占全国大熊猫栖息地面积的70.08%），保护着大熊猫及区域内8000多种野生动植物。建立大熊猫国家公园对于促进人与自然和谐共生、推进美丽中国建设，具有极其重要的意义。

作为普通公民，你在日常生活中就可以采取各种行动或者做出很多选择来保护大熊猫和其他动植物，以及它们赖以生存的自然生态环境。

Over the past century, as a flagship species, umbrella species, the giant panda has played an important role in effectively rescuing and protecting rare and endangered species and promoting biodiversity.

Today, to achieve scientific protection and rational utilization of natural resources, the state approved establishing the Giant Panda National Park. The pilot area for the Giant Panda National Park covers, according to the designs, the Min Mountain Area and Qionglai-Daxiangling Mountain Area in Sichuan province, Qinling Mountain Area in Shaanxi province, and Baishuijiang Area in Gansu province. Over 80 protected areas were reorganized into the Giant Panda National Park, forming a total area of 27,134 km² (accounting for 70.08% of the habitat in China for giant pandas) for 1,631 wild giant pandas (accounting for 87.5% of the total wild population nationwide) as well as more than 8,000 wild flora and fauna species.

As an ordinary citizen, you can also protect pandas and other animals and plants as well as the natural ecological environment by taking various actions or making different choices in your daily life.

日常生活中可以这样做

1. 种植、养殖本地物种，拒绝外来物种；
2. 主动制止破坏环境的行为；
3. 坚决不吃野生动植物，拒绝购买和使用野生动植物及其制品；
4. 尽量减少使用一次性物品，避免购买过度包装的商品；
5. 倡导绿色生活，节约能源，不过度消费；
6. 绿色出行，尽量乘坐公共交通工具；
7. 使用充电电池，注意回收利用普通电池。

Observe the Following Every Day

1 Plant and breed native species and do not introduce foreign species;
2 Actively stop actions that damage the environment and wildlife;
3 Resolutely refuse to eat, buy, and use wildlife and their products;
4 Minimize the use of disposable items and refuse excessively packaged products;
5 Advocate green life and energy saving, and refuse blind and excessive consumption;
6 Advocate green travel and take public transportation as often as possible;
7 Use rechargeable batteries and recycle non-rechargeable batteries.

参观动物园等圈养机构时可以这样做

1. 不向动物投食；
2. 不敲打动物笼舍或玻璃；
3. 不带宠物入园；
4. 不参与动物商业合影；
5. 不追逐动物；
6. 不用奇怪的声音和动作惊吓动物；
7. 拍照时关闭闪光灯。

Observe the Following When Visiting a Zoo and Other Captive Institutions

1 Do not feed the animals;
2 Do not knock on animal cage or glass;
3 Do not bring pets into zoo;
4 Do not participate in commercial animal photography;
5 Do not chase animals;
6 Do not make strange sounds or movements to frighten animals;
7 Turn off your flash when taking pictures.

户外旅行时还可以这样做

1. 不在未经规划的区域开展旅游活动；
2. 不乱扔垃圾或留下任何废弃物；
3. 不要试图靠近、触摸或抓走任何野生动物；
4. 不破坏或带走任何自然物，如花草、石头、树枝、贝壳等；
5. 不要在野外生火，也不要乱扔烟头；
6. 不向你看到的动物吼叫，不干扰野生动物的正常生活；
7. 不要把任何种子、植物或动物带到野外。

Observe the Following When on a Field Trip

1 Do not tour unplanned areas;
2 Do not litter, and take away any garbage;
3 Do not try to approach, touch, or capture any wild animals;
4 Do not harm or take away any natural objects, including flowers, stones, branches, and shells;
5 Do not light a fire or dispose of your cigarette butts in the wild;
6 Do not yell at the animals you see or disturb their normal behavior in the wild;
7 Do not bring any seeds, plants, or animals into the wild.

大熊猫小百科

旗舰物种

指某个物种对生态保护具有特殊号召力和吸引力,可促进社会对物种保护的关注,是地区生态维护的代表物种。这类物种的存亡一般对保持生态过程或食物链的完整性和连续性无严重的影响,但其魅力(外貌或其他特征)赢得了人们的喜爱和关注(如大熊猫、白鳍豚、金丝猴等)。对这类动物的保护容易募集到更多的资金,从而用于保护大规模的生态系统。

伞护种

美国科学家布鲁斯·威尔科克斯于1984年最早提出"伞护种"的概念。伞护种是一个合适的目标物种,这个目标物种的生态环境需求能涵盖其他物种的生态环境需求,对该物种进行保护的同时也为其他物种提供了保护。这种目标物种的生态环境需求应与其他物种的生态环境需求相结合。伞护种可用来确定应被保护的生态环境的类型与面积,常用于自然保护区的规划。

大熊猫就是一个典型的伞护种。保护大熊猫,不仅保护了这一个物种,而且保护了大熊猫栖息地内许许多多的其他物种。

GIANT PANDAS TIDBITS

Flagship Species

It refers to a species that possesses a unique appeal and attraction for public ecological protection, draws public attention to species conservation, and is a representative species for regional ecological maintenance. The survival of such a species will not severely influence maintaining the integrity and continuity of an ecological process or food chain, but its charm (appearance or other features) wins over the public and garners widespread attention. Take the giant panda, Yangtze River Dolphin, and snub-nosed monkey for example. Protecting them can easily raise more funds to protect large-scale ecosystems.

Umbrella Species

It is a term first coined in 1984 by American scientists Bruce Wilcox, who believed that umbrella species is a suitable target species, where its environmental needs covers the environmental needs of many other species and protecting this species provides a conservation umbrella for other species. The environmental needs of this target species should be integrated with that of other species. Umbrella species are often adopted when planning nature reserves to determine the type and area of the living environment to be protected.

The giant panda is the most recognizable umbrella species, and its conservation extends to numerous other species inhabiting its habitat.

致 谢

本书的出版，得到了国家林业和草原局项目课题以及成都市科学技术局科普创作资助项目的资助。这使我能静下心来，把从事科普工作十多年来的知识积累，包括我认为大家对大熊猫不了解的地方，甚至存在误区的地方，通俗易懂地写出来。

感谢我的好朋友西南民族大学文新学院秦丽副教授、成都市花园（国际）小学马佳老师在初稿完成之时给予我的指导和帮助我对书稿所做的修改，她们的建议让书里的内容变得丰富有趣，更加适合小学生阅读。

感谢我的同事侯蓉、黄祥明、吴孔菊、沈富军、杨奎兴、陈欣，他们作为大熊猫研究领域的专家，对书中专业知识的写作给予了指导；感谢我的同事唐亚飞、向波、许萍对书中知识内容提出的修改建议；感谢同事吴樱在英文审校上付出的辛劳；感谢同事徐星煜在图片整理方面给予的帮助。

我希望你和我一样喜欢本书中的图片。感谢所有允许我在本书中免费使用图片的单位及个人。感谢成都大熊猫繁育研究基地、感谢 pandapia、张

Acknowledgements

The book was published thanks to funding from the National Forestry and Grassland Administration and the Chengdu Science and Technology Bureau. Their support allowed me to settle down and present the knowledge that I have accumulated for more than a decade in popular science work, including giant panda information that I think is rarely known, and even misunderstandings, in an easy-to-understand manner.

I would like to extend my sincere gratitude to my friends: Prof. Qin Li from the School of Literature and Journalism of Southwest Minzu University, and teacher Ma Jia from Chengdu Garden (International) Primary School, for providing instructions and supports on revisions when the first draft came out. Their suggestions have made the book more diverse, more appealing, and more suitable for primary school students.

I am also grateful to my colleagues for their kind help. Hou Rong, Huang Xiangming, Wu Kongju, Shen Fujun, Yang Kuixing, and Chen Xin, as giant panda research experts, guided me on writing a professional book; Tang Yafei, Xiang Bo, Xu Ping, offered suggestions on revising book content; Wu Ying toiled on the English revision; and Xu Xingyu helped me collect pictures.

I hope you enjoy the pictures in this book as much as I do. Thanks to all the organizations and individuals that permitted me to use their pictures in the book without a fee: the Chengdu Research Base of Giant Panda Breeding, pandapia, Zhang Zhihe, Cui Kai, Yong Yan'ge, Lan

志和、崔凯、雍严格、兰景超、廖骏、魏辅文、佘轶。

本书的图书设计以及书中大部分插图的绘制，均由薛先良老师及其团队完成。与薛老师的合作源于成都大熊猫博物馆展陈内容的平面设计。薛老师独特的设计风格、对书稿内容的准确理解与诠释，深得我心。在本书经费紧张的情况下，薛老师不计报酬，几易其稿，帮我完成了图书的整体设计工作，让我深受感动。

最后我要感谢一直支持我的家人。感谢我的先生冯继来，在我写作的每晚熬夜陪伴；感谢我的儿子冯梁瑞（小名：小满），在初稿完成时，作为我的第一位读者，他从小学生的角度，对书稿的排版设计、插图风格等都提出了很好的建议。

感谢大家的支持与帮助，祝愿一切都好！

金 双

Jingchao, Liao Jun, Wei Fuwen, She Yi.

    The design book and most illustrations were completed by Mr. Xue Xianliang and his team. My cooperation with Mr. Xue originated from graphic designs for the Chengdu Giant Panda Museum exhibition, and I appreciated Mr. Xue's unique design style and accurate understanding and interpretation of book contents. Under the circumstances of a tight budget, Mr. Xue, waiving remuneration, revised several drafts and helped me complete the overall design for the book, which has deeply moved me.

    Finally, I would like to thank my family that has always supported me. My husband Feng Jilai was with me as I wrote every night; my son Feng Liangrui (nickname: Xiaoman), my first reader, gave me helpful suggestions on content style, layout design, illustration style, etc. from the perspective of a primary school student when the first draft was completed.

    Thank you for your support and help, and I wish you all the best!

Jin Shuang

成都大熊猫繁育研究基地
CHENGDU RESEARCH BASE OF GIANT PANDA BREEDING

　　成都大熊猫繁育研究基地成立于1987年，长期致力于以大熊猫为代表的珍稀濒危动物的保护和研究，并已取得巨大进展和可喜成绩。如果没有公众的参与和支持，仅靠科研人员的力量，拯救这些珍稀濒危动物的目标将很难实现。成都熊猫基地科普教育团队成立于2000年，本书是该团队的诸多努力之一，旨在向公众普及大熊猫科学知识，唤起公众的支持，减少人们对自然环境的破坏，从而缓解大熊猫等珍稀濒危动物受到的生存威胁。

　　The Chengdu Research Base of Giant Panda Breeding has long been committed to the research and conservation of rare and endangered wildlife represented by the giant panda. Since its founding in 1987, the Panda Base has made great progress and strides in these fields, however, we know that without the support and participation of the public we cannot achieve our goal. We know that research alone cannot save wildlife, therefore we established the Conservation Education Department of the Panda Base in 2000. This book is one of our efforts to not only educate the public about the basic biology of giant pandas and their habitat, but also inspire public support to conserve the environment and lessen the threats to the survival of giant pandas and other rare wildlife.